T0200988

THERMOHYDRO-DYNAMIC INSTABILITY IN FLUID-FILM BEARINGS

THERMOHYDRO-DYNAMIC INSTABILITY IN FLUID-FILM BEARINGS

J. K. Wang
Director of Engineering and Product Development
TSC Group Holdings Ltd.
Houston, TX, USA

M. M. Khonsari
Dow Chemical Endowed Chair in Rotating Machinery
Department of Mechanical Engineering
Louisiana State University
Baton Rouge, LA, USA

WILEY

This edition first published 2016
© 2016 John Wiley & Sons, Ltd.

Registered Office
John Wiley & Sons, Ltd, The Atrium, Southern Gate, Chichester, West Sussex, PO19 8SQ,
United Kingdom

For details of our global editorial offices, for customer services and for information about how to apply
for permission to reuse the copyright material in this book please see our website at www.wiley.com.

Library of Congress Cataloging-in-Publication Data

Wang, J. K., author.
  Thermohydrodynamic instability in fluid-film bearings / J. K. Wang, M. M. Khonsari.
    pages cm
  Includes index.
  ISBN 978-0-470-05721-6 (cloth)
  1. Fluid-film bearings–Design and construction.   2. Hydraulics.   3. Fluid dynamics.   I. Khonsari,
Michael M., author.   II. Title.
  TJ1073.5.W36 2015
  621.8′22–dc23
                                                    2015025657

A catalogue record for this book is available from the British Library.

Set in 10/12pt Times by SPi Global, Pondicherry, India
Printed and bound in Singapore by Markono Print Media Pte Ltd

1 2016

Dedicated to:
Yingyu, Jonathan, and Joyce Wang, and Karen, Maxwell, Milton, and Mason Khonsari

Contents

Preface

The importance of rotor-bearing system instability due to oil whirl associated with fluid-film journal bearings has been recognized since its discovery was first reported by Newkirk and Taylor in 1925. While still not well understood, it remains to be a crucially important design consideration in many types of modern rotating machinery.

In this book, we aim to establish the appropriate instability criteria for a rotor-bearing system consisting of a rotor supported by fluid-film journal bearings. In addition to the conventional stability analysis based on linearized stiffness and damping coefficients, the Hopf bifurcation theory (HBT) is employed to describe the nature of important system characteristics that govern the behavior of instability such as the hysteresis phenomenon, stability envelope, dip phenomenon, and subcritical/supercritical bifurcation.

In practice, rotor flexibility, manufacturing imperfections such as residual shaft unbalance, and service-related imperfections (e.g., bearing bushing wear, shaft thermal bow, sag, and any uneven wear or rust) always exist. Moreover, there are operating conditions that require special considerations such as turbulent flows. In this book, the questions on how these factors affect the stability of a rotor-bearing system are answered systematically.

Aside from the running speed, bearing load, and oil grade, there are other operating parameters of a rotor-bearing system affect the system stability. These include, for example, the oil inlet temperature, inlet pressure, and inlet position. The influences of these operating parameters on the stability of rotor-bearing system are also discussed and general design guidelines are provided at the end of each section.

The material presented in this book is largely derived from a series of published research articles by the authors. Here, the concepts are presented in a self-contained and coherent manner for use not only by academic researchers but also by the practicing mechanical engineers and vibration analysts. It is hoped that readers find the presented methodologies and design guidelines useful for treating any rotor-bearing system with any specific set of operating parameters in an effort to improve system performance and guard against failures.

J. K. Wang
M. M. Khonsari

Acknowledgements

The authors wish to express deep appreciation to the permissions from American Society of Mechanical Engineers, SAGA Publication Limited, and Elsevier Limited to re-use some of our published materials.

The authors would like to thank Dr. Brian Hassard from the Department of Mathematics of The State University of New York at Buffalo for his invaluable help and discussions during the development of Hopf bifurcation theory (HBT) for application in rotor-bearing systems presented in this book.

Thanks also go to all the post-doctoral research associates, former Ph.D. and graduate students in the Center for Rotating Machinery (CeRoM) at Louisiana State University. We thank them for their invaluable feedback in the weekly CeRoM internal workshops. Special thanks go to CeRoM associate Mr. Darryl Chauvin, currently with the Global Technology Center of The Dow Chemical Company, who initially developed the test rig which was used in the experiments presented in this book.

1

Fundamentals of Hydrodynamic Bearings

Hydrodynamic (fluid film) bearings are used extensively in different kinds of rotating machinery in the industry. Their performance is of utmost importance in chemical, petrochemical, automotive, power generation, oil and gas, aerospace turbo-machinery, and many other process industries around the globe.

Hydrodynamic bearings are generally classified into two broad categories: journal bearings (also called sleeve bearings) and thrust bearings (also called slider bearings). In this book, we exclusively focus our attention on journal bearings.

Figure 1.1a shows a schematic illustration of a rotor bearing system, which consists of a shaft with a central disk symmetrically supported by two identical journal bearings at both ends. Figure 1.1b shows the geometry and system coordinates of the journal rotating in one of the two identical journal bearings. To easily identify the bearing's physical wedge effect and annotate the multiple parameters of a rotor bearing system, the clearance between the journal and the bearing bushing is exaggerated. θ is the circumferential coordinate starting from the line going through the centers of the bearing bushing and the rotor journal. ϕ is defined as the system attitude angle. e is the rotor journal center eccentricity from the center of the bearing bushing. W represents the vertical load imposed on the shaft and supported by the bearing. p is the hydrodynamic pressure applied by the thin fluid film onto the journal surface. f is the hydrodynamic force obtained by integrating the hydrodynamic pressure p generated around the journal circumference.

Thermohydrodynamic Instability in Fluid-Film Bearings, First Edition.
J. K. Wang and M. M. Khonsari.
© 2016 John Wiley & Sons, Ltd. Published 2016 by John Wiley & Sons, Ltd.

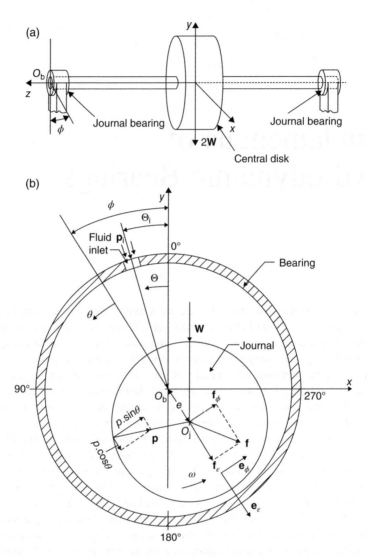

Figure 1.1 (a) Model of a rotor supported by two identical journal bearings; (b) geometry and system coordinates of a journal rotating in a fluid film journal bearing

In most cases, except in a floating ring configuration, the bearing bushing is fixed and the rotor rotates at the speed of ω inside the bearing bushing. In Figure 1.1, the journal center position O_j is described as (e, ϕ) relative to the center O_b of the fixed journal bearing bushing.

Radial clearance C is defined as the clearance between the bearing and the rotor journal (i.e., $C = R_b - R_j$, where R_b is the inside radius of the bearing bushing and R_j

is the radius of the rotor journal). In terms of this radial clearance, the journal center eccentricity from the bearing center can be normalized as $\varepsilon = e/C$. The dimensionless parameter ε is called eccentricity ratio. Due to the physical constraint of the bearing bushing, the rotor journal must be designed to operate inside of the bearing bushing, that is, $0 \le \varepsilon \le 1$. Therefore, the journal center position O_j within the fluid film journal bearing can be redefined as $(C\varepsilon, \phi)$. When $\varepsilon = 0$, the center of the shaft O_j coincides with the center of the bearing bushing O_b and the fluid film bearing is theoretically incapable of generating hydrodynamic pressure by wedge effect and its corresponding load-carrying capacity is nil. When $\varepsilon = 1$, the shaft comes into intimate contact with the inner surface of the bushing, and depending on the operating speed, bearing failure becomes imminent due to the physical rubbing between the shaft and the bushing.

Based on the above physics, the important concept of rotor bearing clearance circle is introduced to easily describe the rotor journal position within any hydrodynamic journal bearing. Figure 1.2a shows the rotor bearing clearance circle in both polar and Cartesian coordinate systems. The radius of the clearance circle is equal to the radial clearance C defined earlier and the center of the clearance circle is the bearing center O_b. The journal center O_j is always either within or on the clearance circle. In other words, it will never go beyond the clearance circle due to the physical constraint of bearing bushing. Figure 1.2b shows the dimensionless rotor bearing clearance circle in both polar and Cartesian coordinate systems.

The fundamental equation that governs the pressure distribution in a hydrodynamic bearing was first introduced by Osborne Reynolds in 1886. In this chapter, we begin by describing the Reynolds equation and provide closed-form analytical solutions for two simplified extreme cases commonly known as the short and long bearing solutions. At the end, a brief discussion will be provided to address the numerical methods to solve the Reynolds equation for finite-length journal bearings.

1.1 Reynolds Equation

The Reynolds equation assuming that thin-film lubrication theory holds for a perfectly aligned journal bearing system lubricated with an incompressible Newtonian fluid is given by Equation 1.1.

$$\frac{\partial}{R^2 \partial\theta}\left(G_\theta \frac{h^3}{\mu}\frac{\partial p}{\partial\theta}\right) + \frac{\partial}{\partial z}\left(G_z \frac{h^3}{\mu}\frac{\partial p}{\partial z}\right) = \frac{\omega}{2}\frac{\partial h}{\partial\theta} + \frac{\partial h}{\partial t} \qquad (1.1)$$

where z is the axial coordinate with the origin at the mid-width of the journal bearing.

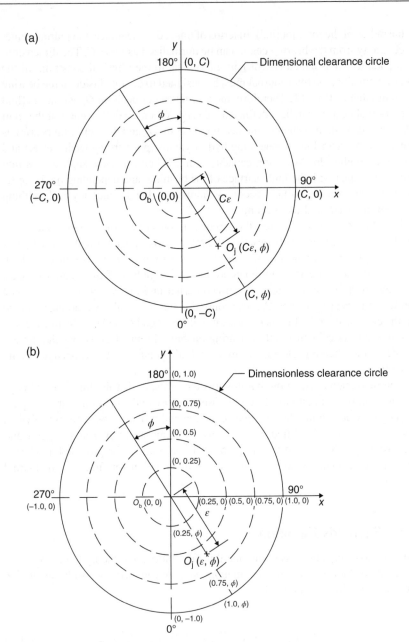

Figure 1.2 (a) Dimensional and (b) dimensionless rotor bearing clearance circles

Detailed derivation of the Reynolds equation is available in tribology textbooks (see for example, Khonsari and Booser, 2008). In Equation 1.1, θ is the circumferential coordinate and z is the axial coordinate perpendicular to the paper in Figure 1.1, R is the journal radius, μ is the fluid viscosity, and the fluid film thickness h is given by Equation 1.2. The parameters G_θ and G_z are the turbulent coefficients given by Equations 1.3 and 1.4 (See Hashimoto and Wada (1982) and Hashimoto et al. (1987)).

$$h = C(1 + \varepsilon \cos\theta) \tag{1.2}$$

$$G_\theta = \frac{1}{12(a_\theta + b_\theta \varepsilon \cos\theta)} \tag{1.3}$$

$$G_z = \frac{1}{12(a_z + b_z \varepsilon \cos\theta)} \tag{1.4}$$

where $a_\theta = 1 + 0.00069\text{Re}^{0.95}$, $a_z = 1 + 0.00069\text{Re}^{0.88}$, $b_\theta = 0.00066\text{Re}^{0.95}$, $b_z = 0.00061\text{Re}^{0.88}$, and $\text{Re} = \rho R\omega C/\mu$ is the Reynolds number. The turbulent coefficients G_θ and G_z given by Equations 1.3 and 1.4 agree well with those given by Ng and Pan (1965).

On the left-hand side of Reynolds Equation 1.1, the first term $\left(\frac{\partial}{R^2\partial\theta}\left(G_\theta \frac{h^3}{\mu}\frac{\partial p}{\partial\theta}\right)\right)$ is the pressure-induced flow in the circumferential direction while the second term $\left(\frac{\partial}{\partial z}\left(G_z \frac{h^3}{\mu}\frac{\partial p}{\partial z}\right)\right)$ is the pressure-induced flow in the axial direction. On the right-hand side, the first term $\left(\frac{\omega}{2}\frac{\partial h}{\partial\theta}\right)$ is the physical wedge effect in the circumferential direction between the bearing bushing and the rotor journal, and the second term $(\partial h/\partial t)$ is the normal squeeze action of the fluid film in the radial direction.

Under the simplified isothermal assumption and neglecting the pressure influence on the fluid viscosity (i.e., constant fluid viscosity throughout the fluid film), the Reynolds Equation 1.1 can be simplified to

$$\frac{\partial}{R^2\partial\theta}\left(G_\theta h^3 \frac{\partial p}{\partial\theta}\right) + \frac{\partial}{\partial z}\left(G_z h^3 \frac{\partial p}{\partial z}\right) = \frac{\omega\mu}{2}\frac{\partial h}{\partial\theta} + \mu\frac{\partial h}{\partial t} \tag{1.5}$$

For a steady-state fluid film, the fluid film thickness h is not a function of time, that is, $\partial h/\partial t = 0$. Then, the Reynolds equation can be further reduced to

$$\frac{\partial}{R^2\partial\theta}\left(G_\theta h^3 \frac{\partial p}{\partial\theta}\right) + \frac{\partial}{\partial z}\left(G_z h^3 \frac{\partial p}{\partial z}\right) = \frac{\omega\mu}{2}\frac{\partial h}{\partial\theta} \tag{1.6}$$

while for laminar flow ($a_\theta = 1, a_z = 1, b_\theta = 0, b_z = 0$), the simplified Reynolds equation (Eq. 1.1) can be rewritten as

$$\frac{\partial}{R^2\partial\theta}\left(\frac{h^3}{\mu}\frac{\partial p}{\partial\theta}\right) + \frac{\partial}{\partial z}\left(\frac{h^3}{\mu}\frac{\partial p}{\partial z}\right) = 6\omega\frac{\partial h}{\partial\theta} + 12\frac{\partial h}{\partial t} \qquad (1.7)$$

Therefore, for a rotor bearing system with steady-state and laminar fluid film, Equation 1.8 presents the further reduced but commonly used Reynolds equation.

$$\frac{\partial}{R^2\partial\theta}\left(\frac{h^3}{\mu}\frac{\partial p}{\partial\theta}\right) + \frac{\partial}{\partial z}\left(\frac{h^3}{\mu}\frac{\partial p}{\partial z}\right) = 6\omega\frac{\partial h}{\partial\theta} \qquad (1.8)$$

Reynolds Equation 1.1 is a time-dependent second-order partial differential equation. To predict the pressure distribution through solving the Reynolds equation, in addition to the initial condition, four boundary conditions are needed in terms of the geometrical parameters θ and z. For steady-state Reynolds equations such as Equations 1.6 and 1.8, only the four boundary conditions are needed to define the pressure distribution.

1.1.1   Boundary Conditions for Reynolds Equation

In most hydrodynamic bearing applications, the fluid lubricant flows out of the bearing at ambient pressure. In other words, the gauge pressure at the geometrical boundary is equal to 0. Inside the bearings, since a conventional fluid lubricant cannot withstand negative pressure, it cavitates if the liquid pressure falls below the atmospheric pressure.

Depending on how to define and handle the cavitation region, there are three classical types of boundary conditions: full-Sommerfeld boundary conditions (cavitation is fully neglected and $p = 0$ when $\theta = 2\pi$), half-Sommerfeld boundary conditions (also called Gümbel boundary conditions, i.e., $p = 0$ when $180° \leq \theta \leq 360°$), and Reynolds boundary conditions (also called Swift–Stieber boundary conditions, i.e., both pressure and pressure gradient approach 0 where cavitation begins). All three classical types of boundary conditions assume that the fluid film starts at $\theta = 0$. The detailed definitions of these boundary conditions will be introduced in the related chapters that follow. For further reference, Khonsari and Booser (2008) have given a complete summary of these boundary conditions on both their implications and limitations. In recent years, by combining the Reynolds boundary condition with some new experimental findings on when and how the fluid film starts, a more complete type of boundary conditions (Reynolds–Floberg–Jakobsson or RFJ boundary conditions) has been derived and applied successfully into different applications (Wang and Khonsari, 2008). The RFJ boundary conditions will be discussed in Section 1.3.

As Equations 1.1, 1.5–1.8 read, even for the most simplified two-dimensional Reynolds equation—which is still a nonlinear partial differential equation—a closed-form analytical solution is practically impossible.

On the other hand, based on the physical implications of different applications, two kinds of extreme-condition approximation of the Reynolds equation have been well developed and applied widely to predict the bearing performance analytically. One of them is called the infinitely short bearing theory (often called the short bearing theory) for the application of bearing length-over-diameter ratio far less than 1; the other approximation is the infinitely long bearing theory (often called the long bearing theory) for the application of bearing length-over-diameter ratio far more than 1.

Generally speaking, to obtain sufficiently accurate results, the short bearing theory is often applied to bearings with length-over-diameter ratio up to 0.5 and the infinitely long bearing theory is recommended for bearings with length-over-diameter ratio of 2.0 or greater. The bearing having length-over-diameter ratio more than 0.5 while not exceeding 2.0 is called finite bearing. For finite bearings with length-over-diameter ratio more than 0.5 while not exceeding 1.0, the short bearing theory still could render a reasonable approximation of the bearing performance. If the finite bearing length-over-diameter ratio is more than 1.0, the long bearing theory might be a reasonable approximation, particularly if one is interested in trends. However, if an accurate prediction of the bearing performance is desired, then the full Reynolds equation should be treated with an appropriate numerical solution.

1.1.2 Short Bearing Approximation

For short journal bearings, the second term (side leakage in the axial direction) in the Reynolds Equations 1.1, 1.5–1.8 is so dominant that the first term (the pressure-induced flow in the circumferential direction) can be neglected to obtain an analytical solution to the Reynolds equations. The side leakage controls the fluid film pressure distribution and then the bearing loading capacity. Section 1.2 will show the simplified Reynolds equation for short bearings under certain commonly used boundary conditions.

1.1.3 Long Bearing Approximation

For long journal bearings, side leakage and fluid pressure gradient in the axial direction are negligible (i.e., $\partial p / \partial z \approx 0$). This implies that the second term (side leakage in the axial direction) in the Reynolds Equations 1.1, 1.5–1.8 can be dropped and it becomes possible to obtain an analytical solution to the Reynolds equations. Section 1.3 will show the detailed derivation of an analytical solution for long bearings under certain boundary conditions.

1.2   Short Bearing Theory

1.2.1   Analytical Pressure Distribution

In infinitely short journal bearings, side leakage controls the fluid film pressure distribution and the bearing load-carrying capacity. Because of the dominant pressure-induced side leakage in the axial direction, the pressure-induced flow in the circumferential direction (i.e., the partial differentials of the pressure p in terms of θ on the left-hand side of Reynolds Equation 1.1) is neglected to obtain an analytical solution to the Reynolds equations. Assuming constant fluid viscosity throughout the fluid film, the Reynolds equation for infinitely short bearings including the turbulent effects can be further reduced from Equations 1.5 to 1.9.

$$G_z \frac{\partial^2 p}{\partial z^2} = \frac{\mu\omega}{2h^3}\frac{\partial h}{\partial \theta} + \frac{\mu}{h^3}\frac{\partial h}{\partial t} \tag{1.9}$$

The boundary conditions of the fluid film in the axial directions are given by Equations 1.10 and 1.11.

$$\left.\frac{\partial p}{\partial z}\right|_{z=0} = 0 \tag{1.10}$$

$$p\big|_{z=\pm(L/2)} = 0 \tag{1.11}$$

where L is the bearing length.

Integrating the Reynolds Equation 1.9 twice and substituting the above boundary conditions yields the following expression for the pressure distribution of the fluid film formed around the journal surface:

$$p = \frac{3\mu(a_z + b_z\varepsilon\cos\theta)}{h^3}\left(\omega\frac{\partial h}{\partial \theta} + 2\frac{\partial h}{\partial t}\right)\left(z^2 - \frac{L^2}{4}\right) \tag{1.12}$$

where a_z and b_z are defined in Section 1.1.

Substituting Equation 1.2 into Equation 1.12, the expression for hydrodynamic pressure distribution becomes (Wang and Khonsari, 2006)

$$p = \frac{3\mu}{C^2}\left(z^2 - \frac{L^2}{4}\right)$$

$$\frac{2a_z\dot{\varepsilon}\cos\theta - a_z(\omega+2\dot{\theta})\varepsilon\sin\theta + 2b_z\varepsilon\dot{\varepsilon}\cos^2\theta - b_z(\omega+2\dot{\theta})\varepsilon^2\sin\theta\cos\theta}{(1+\varepsilon\cos\theta)^3} \tag{1.13}$$

where "." represents d/dt.

Referring to Figure 1.1, the position of the journal center can be represented as O_j $(C\varepsilon, \phi)$, where ϕ denotes the attitude angle. Note that as ϕ changes, so does the reference line for θ. Increasing ϕ will result in decreasing θ (Lund, 1966); so,

$$\frac{d\phi}{dt} = -\frac{d\theta}{dt} \quad \text{i.e.} \quad \dot{\theta} = -\dot{\phi} \tag{1.14}$$

With the aid of Equation 1.14, the expression for the pressure distribution, Equation 1.13, can be rewritten as (Wang and Khonsari, 2006)

$$p = \frac{3\mu}{C^2}\left(z^2 - \frac{L^2}{4}\right)$$

$$\frac{2a_z\dot{\varepsilon}\cos\theta - a_z(\omega - 2\dot{\phi})\varepsilon\sin\theta + 2b_z\varepsilon\dot{\varepsilon}\cos^2\theta - b_z(\omega - 2\dot{\phi})\varepsilon^2\sin\theta\cos\theta}{(1 + \varepsilon\cos\theta)^3} \tag{1.15}$$

If the fluid film has a laminar flow ($a_z = 1$, $b_z = 0$), the pressure distribution further simplifies to the following expression (Wang and Khonsari, 2006):

$$p = \frac{3\mu}{C^2}\left(z^2 - \frac{L^2}{4}\right)\frac{2\dot{\varepsilon}\cos\theta - (\omega - 2\dot{\phi})\varepsilon\sin\theta}{(1 + \varepsilon\cos\theta)^3} \tag{1.16}$$

1.2.2   Hydrodynamic Fluid Force

Figure 1.3 illustrates the radial and tangential components of the fluid force exerted on the rotor journal surface. By definition, the fluid force components in the radial (subscript ε) and tangential (subscript ϕ) directions can be obtained by integrating the pressure distribution around the rotor journal surface:

$$f_\varepsilon = R\int_{-L/2}^{L/2}\int_0^{2\pi} p\cos\theta\, d\theta\, dz \tag{1.17}$$

$$f_\phi = R\int_{-L/2}^{L/2}\int_0^{2\pi} p\sin\theta\, d\theta\, dz \tag{1.18}$$

Substituting Equation 1.15 into Equations 1.17 and 1.18, applying the half Sommerfeld boundary condition where the negative hydrodynamic fluid pressure at the geometrically diverging region is assumed to be negligible (i.e., $p = 0$ when

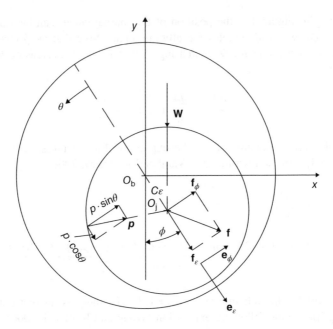

Figure 1.3   Radial and tangential components of the fluid force exerted on the journal (Wang and Khonsari, 2006)

$180° \leq \theta \leq 360°$), and utilizing the integrals given in the appendix A by Wang and Khonsari (2006), the following expressions for the hydrodynamic fluid force components yield (Wang and Khonsari, 2006)

$$
\begin{aligned}
f_\varepsilon = -\frac{\mu R L^3}{2C^2} &\left\{ (\omega - 2\dot{\phi}) \left[\frac{2(a_z - 2b_z)\varepsilon^2 + 2b_z}{(1-\varepsilon^2)^2} - \frac{b_z}{\varepsilon} \ln\left(\frac{1+\varepsilon}{1-\varepsilon}\right) \right] \right. \\
&\left. + \pi \dot{\varepsilon} \left[\frac{a_z + 5b_z + 2(a_z - 3b_z)\varepsilon^2 - (2b_z/\varepsilon^2)}{(1-\varepsilon^2)^{5/2}} + \frac{2b_z}{\varepsilon^2} \right] \right\}
\end{aligned}
\tag{1.19}
$$

$$
\begin{aligned}
f_\phi = \frac{\mu R L^3}{2C^2 \varepsilon} &\left\{ \pi(\omega - 2\dot{\phi}) \left[\frac{2b_z + (a_z - 3b_z)\varepsilon^2}{2(1-\varepsilon^2)^{3/2}} - b_z \right] \right. \\
&\left. + 4\dot{\varepsilon} \left[\frac{b_z + (a_z - 2b_z)\varepsilon^2}{(1-\varepsilon^2)^2} - \frac{b_z}{2\varepsilon} \ln\left(\frac{1+\varepsilon}{1-\varepsilon}\right) \right] \right\}
\end{aligned}
\tag{1.20}
$$

where "." represents d/dt.

Normalizing the time t parameter by substituting $\bar{t} = \omega t$, the dimensionless radial and tangential components of the hydrodynamic fluid force in the journal bearing become

$$\bar{f}_\varepsilon = \frac{f_\varepsilon}{mC\omega^2} = -\frac{\mu R L^3}{2mC^3\omega}\left\{ (1-2\dot{\phi})\left[\frac{2(a_z-2b_z)\varepsilon^2+2b_z}{(1-\varepsilon^2)^2} - \frac{b_z}{\varepsilon}\ln\left(\frac{1+\varepsilon}{1-\varepsilon}\right)\right]\right.$$

$$\left.+\pi\dot{\varepsilon}\left[\frac{a_z+5b_z+2(a_z-3b_z)\varepsilon^2-(2b_z/\varepsilon^2)}{(1-\varepsilon^2)^{5/2}} + \frac{2b_z}{\varepsilon^2}\right]\right\} \qquad (1.21)$$

$$\bar{f}_\phi = \frac{f_\phi}{mC\omega^2} = \frac{\mu R L^3}{2mC^3\omega\varepsilon}\left\{ \pi(1-2\dot{\phi})\left[\frac{2b_z+(a_z-3b_z)\varepsilon^2}{2(1-\varepsilon^2)^{3/2}} - b_z\right]\right.$$

$$\left.+4\dot{\varepsilon}\left[\frac{b_z+(a_z-2b_z)\varepsilon^2}{(1-\varepsilon^2)^2} - \frac{b_z}{2\varepsilon}\ln\left(\frac{1+\varepsilon}{1-\varepsilon}\right)\right]\right\} \qquad (1.22)$$

where "." represents $d/\omega dt$.

Under laminar flow condition ($a_z = 1$ and $b_z = 0$), the hydrodynamic fluid force components (Eqs. 1.19 and 1.20) are simplified as

$$f_\varepsilon = -\frac{\mu R L^3}{2C^2}\left[\frac{2(\omega-2\dot{\phi})\varepsilon^2}{(1-\varepsilon^2)^2} + \frac{\pi\dot{\varepsilon}(1+2\varepsilon^2)}{(1-\varepsilon^2)^{5/2}}\right] \qquad (1.23)$$

$$f_\phi = \frac{\mu R L^3}{2C^2}\left[\frac{\pi(\omega-2\dot{\phi})\varepsilon}{2(1-\varepsilon^2)^{3/2}} + \frac{4\varepsilon\dot{\varepsilon}}{(1-\varepsilon^2)^2}\right] \qquad (1.24)$$

where "." represents d/dt.

Equations 1.25 and 1.26 show the dimensionless hydrodynamic fluid force components for laminar flow applications.

$$\bar{f}_\varepsilon = \frac{f_\varepsilon}{mC\omega^2} = -\frac{\mu R L^3}{2mC^3\omega}\left[\frac{2(1-2\dot{\phi})\varepsilon^2}{(1-\varepsilon^2)^2} + \frac{\pi\dot{\varepsilon}(1+2\varepsilon^2)}{(1-\varepsilon^2)^{5/2}}\right] \qquad (1.25)$$

$$\bar{f}_\phi = \frac{f_\phi}{mC\omega^2} = \frac{\mu R L^3}{2mC^3\omega}\left[\frac{\pi(1-2\dot{\phi})\varepsilon}{2(1-\varepsilon^2)^{3/2}} + \frac{4\varepsilon\dot{\varepsilon}}{(1-\varepsilon^2)^2}\right] \qquad (1.26)$$

where "." represents $d/\omega dt$.

1.2.3 Static Performance of Short Journal Bearings

Assuming that the rate of speed change is zero ($\dot{\varepsilon} = \dot{\phi} = 0$), the fluid pressure, integrated fluid force components, attitude angle, and the relation between eccentricity

Table 1.1 Comparison of static performance based on different types of boundary condition

Half-Sommerfeld (Gümbel) boundary condition	Full-Sommerfeld boundary condition
Pressure:	Pressure:
$$p=\frac{3\mu\omega\varepsilon\sin\theta}{C^2(1+\varepsilon\cos\theta)^3}\left(\frac{L^2}{4}-z^2\right)$$ when $0°\le\theta\le180°$; $p=0$ when $180°<\theta<360°$. For both cases, $-L/2\le z\le L/2$	$$p=\frac{3\mu\omega\varepsilon\sin\theta}{C^2(1+\varepsilon\cos\theta)^3}\left(\frac{L^2}{4}-z^2\right)$$ when $0°\le\theta<360°$ and $-L/2\le z\le L/2$.
Fluid force components:	Fluid force components:
$$f_\varepsilon=-\frac{\mu RL^3\omega\varepsilon^2}{C^2(1-\varepsilon^2)^2}$$ $$f_\phi=\frac{\pi\mu RL^3\omega\varepsilon}{4C^2(1-\varepsilon^2)^{3/2}}$$	$f_\varepsilon=0$ $$f_\phi=\frac{\pi\mu RL^3\omega\varepsilon}{2C^2(1-\varepsilon^2)^{3/2}}$$
Relation between Sommerfeld number and eccentricity ratio:	Relation between Sommerfeld number and eccentricity ratio:
$$S=\frac{4R^2(1-\varepsilon^2)^2}{\pi L^2\varepsilon\sqrt{16\varepsilon^2+\pi^2(1-\varepsilon^2)}}$$	$$S=\frac{2R^2(1-\varepsilon^2)^{3/2}}{\pi^2L^2\varepsilon}$$
Attitude angle:	Constant attitude angle:
$$\phi=\tan^{-1}\left(\frac{\pi\sqrt{1-\varepsilon^2}}{4\varepsilon}\right)$$	$$\phi=\frac{\pi}{2}$$

ratio ε and Sommerfeld number S ($S=\mu\omega LR^3/(\pi WC^2)$) can be derived using Equations 1.16 and 1.23–1.24 for laminar flow applications and the force equilibrium between the hydrodynamic fluid force $\mathbf{f}$ and the externally applied load $\mathbf{W}$ on the journal ($W=\sqrt{f_\varepsilon^2+f_\phi^2}$ and $\tan\phi=-f_\phi/f_\varepsilon$). Table 1.1 shows the comparison between the results derived by applying half-Sommerfeld boundary condition and those based on full-Sommerfeld boundary condition.

The difference between full-Sommerfeld and half-Sommerfeld boundary conditions is in the treatment of the hydrodynamic pressure in the geometrically divergent region of flow. With the half-Sommerfeld boundary condition, the negative fluid pressure p is set to $p=0$ when $180°<\theta<360°$ while with Full-Sommerfeld boundary condition the negative fluid pressure is assumed to remain intact. Under the full-Sommerfeld boundary condition, the skew-symmetric hydrodynamic pressure distribution in the whole circumferential coordinate ($0° \leq \theta < 360°$) yields a constant attitude angle of $\phi = \pi/2$. Numerous studies have proven that cavitation often exists at the geometrically diverging region and the half-Sommerfeld boundary condition setting the negative pressure to zero actually renders a more accurate analytical solution to Reynolds equation, especially for short journal bearings.

1.3   Long Bearing Theory

With identical bearing material of the same PV limit (contact pressure × surface velocity), long jounral bearings normally have a higher load-bearing capacity than short journal bearings. Proper distribution of lubricant requires special consideration, particularly for long journal bearings to function. Often an axial groove is designed to distribute the lubricant over the entire length of the journal to improve the lubrication condition and control the temperature field. The location of the fluid inlet and the associated fluid supply pressure can have a pronounced influence on the bearing performance. For journal bearings with length-over-diameter greater than 1, the infinitely long bearing theory should be used to obtain an analytical solution to the Reynolds Equations 1.1, 1.5–1.8. In addtion, due to its physical implications and the oversimplication in the circumferential direction of the full Reynolds equation, the short bearing theory described in Section 1.2 cannot be used to evaluate the axial groove effects. To obtain an analytical solution of the governing Reynolds equation including the axial groove effects, the long bearing theory is the only option.

1.3.1   Analytical Pressure Distribution of Long Journal Bearings

Figure 1.4 shows the geometry and system coordinates used in long journal bearings. An absolute circumferential coordinate Θ is introduced to facilitate a comparison of the bearing performance under different fluid inlet positions and inlet pressures. It is measured from the upper load line, that is, $\Theta = \theta + \phi$, where θ is the circumferential coordinate starting from the line of the centers of bearing bushing and rotor journal and ϕ is the attitude angle.

Theoretically in infinitely long journal bearings, there is no side leakage and the fluid pressure gradient in the axial direction is zero (i.e. $\partial p / \partial z = 0$). Assuming constant fluid viscosity throughout the fluid film, the Reynolds Equation 1.5 for an infinitely long journal bearing is simplified to

$$\frac{\partial}{\partial \theta}\left(G_\theta h^3 \frac{\partial p}{\partial \theta}\right) = \frac{\mu \omega R^2}{2}\frac{\partial h}{\partial \theta} + \mu R^2 \frac{\partial h}{\partial t} \qquad (1.27)$$

where $h = C(1 + \varepsilon \cos \theta)$, μ is constant throughout the fluid film, and the turbulent coefficient G_θ is given by Equation 1.3.

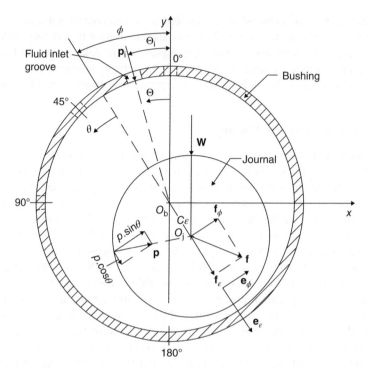

Figure 1.4 Geometry and system coordinates used in long journal bearing. From Wang and Khonsarı (2008) © Elsevier Limited.

If the flow is laminar ($a_z = 1$, $b_z = 0$), the Reynolds equation for a long bearing can be further simplified to (Wang and Khonsari, 2008)

$$\frac{\partial}{\partial \theta}\left(\frac{h^3 \partial p}{\partial \theta}\right) = 6\mu\omega R^2 \frac{\partial h}{\partial \theta} + 12\mu R^2 \frac{\partial h}{\partial t} \qquad (1.28)$$

The boundary conditions are

$$p = 0 \quad \text{at} \quad \theta = \theta_s \qquad (1.29)$$

$$p = 0 \quad \text{at} \quad \theta = \theta_c, \qquad (1.30)$$

$$\frac{\partial p}{\partial \theta} = 0 \quad \text{at} \quad \theta = \theta_c \qquad (1.31)$$

where θ_s is the fluid pressure starting position and θ_c is the circumferential location where the fluid film ruptures; that is, cavitation begins. According to the Reynolds boundary condition, both the pressure and its gradient are equal to zero at θ_c.

Equations 1.30 and 1.31 reflect the Reynolds boundary conditions. According to the Floberg–Jakobsson boundary condition, the pressure at the fluid pressure starting position θ_s is zero while, to satisfy the fluid flow continuity in journal bearing, the pressure gradient $\partial p / \partial \theta$ at the fluid pressure starting position θ_s must be nonnegative (Jakobsson and Floberg, 1957), that is,

$$\frac{\partial p}{\partial \theta} \geq 0 \quad \text{at} \quad \theta = \theta_s \tag{1.32}$$

Since the fluid pressure at the fluid inlet position is always maintained at a specified supply pressure p_i (Zhang, 1989), the following equation introduces an extra boundary condition at the fluid inlet position θ_i.

$$p = p_i \quad \text{at} \quad \theta = \theta_i \tag{1.33}$$

where $\theta_i = \Theta_i - \phi$.

Since $h = C(1 + \varepsilon \cos\theta)$ as given in Equation 1.2, the Reynolds Equation 1.28 for long journal bearing can be rewritten as

$$\frac{\partial}{\partial \theta}\left(\frac{C^3(1+\varepsilon\cos\theta)^3 \partial p}{\partial \theta}\right) = -6\mu\omega CR^2 \varepsilon \sin\theta + 12\mu CR^2 \left(\dot{\varepsilon}\cos\theta - \dot{\theta}\varepsilon\sin\theta\right) \tag{1.34}$$

where "." represents d/dt.

Further, substituting Equation 1.14 into Equation 1.34 and rearranging the equation yields

$$\frac{d}{d\theta}\left(\frac{C^3(1+\varepsilon\cos\theta)^3 dp}{d\theta}\right) = 6\mu CR^2 \varepsilon \left(2\dot{\phi} - \omega\right)\sin\theta + 12\mu CR^2 \dot{\varepsilon}\cos\theta \tag{1.35}$$

Integrating Reynolds Equation 1.35 twice with respect to θ and applying the boundary conditions given by Equations 1.31 and 1.33 (see Appendix A for detailed derivation) yields the following solution for the hydrodynamic pressure distribution around the journal circumference (Wang and Khonsari, 2008):

$$
\begin{aligned}
p = p_i + \frac{3\mu R^2}{C^2(1-\varepsilon^2)^2} \Bigg\{ & \frac{\left(\omega - 2\dot{\phi}\right)\varepsilon\left(1-\varepsilon^2\right)^{0.5}}{(1-\varepsilon\cos\alpha_c)} [2(\sin\alpha - \sin\alpha_i)(1 + \varepsilon\cos\alpha_c) \\
& + (2\cos\alpha_c + \varepsilon)(\alpha_i - \alpha) - \varepsilon(\sin\alpha\cos\alpha - \sin\alpha_i\cos\alpha_i)] \\
& + \dot{\varepsilon}\frac{2\sin\alpha_c}{(1-\varepsilon\cos\alpha_c)}\Bigg[\frac{(2 - \varepsilon\cos\alpha - \varepsilon\cos\alpha_i)(\cos\alpha_i - \cos\alpha)(1 - \varepsilon\cos\alpha_c)}{\sin\alpha_c} \\
& - (2 + \varepsilon^2)(\alpha - \alpha_i) + 4\varepsilon(\sin\alpha - \sin\alpha_i) - \varepsilon^2(\sin\alpha\cos\alpha - \sin\alpha_i\cos\alpha_i)\Bigg]\Bigg\}
\end{aligned}
$$

$$\tag{1.36}$$

The dynamic pressure given by Equation 1.36 is valid for $\alpha_s \leq \alpha \leq \alpha_c$ since the fluid film pressure only exists in the range of $\theta_s \leq \theta \leq \theta_c$ per definition. Beyond this range, the fluid film pressure is zero. The relations between α_s and θ_s, α_c and θ_c, and α_i and θ_i are defined by Equations A.5–A.7, respectively.

By applying the boundary conditions described by Equations 1.29 and 1.30 to determine the parameters in Equation 1.36, the fluid pressure starting position θ_s and the starting position of cavitation θ_c are given by Equations 1.37 and 1.38 (Wang and Khonsari, 2008).

$$
\begin{aligned}
&\left(\omega - 2\dot{\phi}\right)\varepsilon\left(1-\varepsilon^2\right)^{0.5}\left[\left(2+\varepsilon\cos\alpha_c\right)\left(\sin\alpha_c - \sin\alpha_i\right) - \varepsilon\sin\alpha_i\left(\cos\alpha_c - \cos\alpha_i\right)\right. \\
&+\left(2\cos\alpha_c + \varepsilon\right)\left(\alpha_i - \alpha_c\right)\Big] + 2\dot{\varepsilon}\left[\left(2\cos\alpha_i - 2\cos\alpha_c + \varepsilon\sin^2\alpha_i\right)\left(1-\varepsilon\cos\alpha_c\right)\right. \\
&\left.-\varepsilon\sin\alpha_i\sin\alpha_c\left(4-\varepsilon\cos\alpha_i\right) - \left(2+\varepsilon^2\right)\sin\alpha_c\left(\alpha_c - \alpha_i\right) + 3\varepsilon\sin^2\alpha_c\right] \\
&+\frac{C^2 p_i}{3\mu R^2}\left(1-\varepsilon^2\right)^2\left(1-\varepsilon\cos\alpha_c\right) = 0
\end{aligned}
\tag{1.37}
$$

$$
\begin{aligned}
&\left(\omega - 2\dot{\phi}\right)\varepsilon\left(1-\varepsilon^2\right)^{0.5}\left[2\left(\sin\alpha_s - \sin\alpha_i\right)\left(1+\varepsilon\cos\alpha_c\right) + \left(2\cos\alpha_c + \varepsilon\right)\left(\alpha_i - \alpha_s\right)\right. \\
&-\varepsilon\left(\sin\alpha_s\cos\alpha_s - \sin\alpha_i\cos\alpha_i\right)\Big] + 2\dot{\varepsilon}\left[4\varepsilon\sin\alpha_c\left(\sin\alpha_s - \sin\alpha_i\right)\right. \\
&+\left(2-\varepsilon\cos\alpha_s - \varepsilon\cos\alpha_i\right)\left(\cos\alpha_i - \cos\alpha_s\right)\left(1-\varepsilon\cos\alpha_c\right) - \left(2+\varepsilon^2\right)\sin\alpha_c\left(\alpha_s - \alpha_i\right) \\
&\left.-\varepsilon^2\sin\alpha_c\left(\sin\alpha_s\cos\alpha_s - \sin\alpha_i\cos\alpha_i\right)\right] + \frac{C^2 p_i}{3\mu R^2}\left(1-\varepsilon^2\right)^2\left(1-\varepsilon\cos\alpha_c\right) = 0
\end{aligned}
$$

$$
\tag{1.38}
$$

Equations 1.37 and 1.38 show that both α_c and α_s are a function of ε, $\dot{\varepsilon}$, $\dot{\phi}$, and α_i. The parameter α_i is a function of ε and ϕ with a specified fluid inlet position Θ_i. Upon determining α_c by solving Equation 1.37 while $\pi < \alpha_c \leq 2\pi + \alpha_i$ with α_i given by Equation A.7, α_s can be predicted using Equation 1.38 while $(\alpha_c - 2\pi) \leq \alpha_s \leq \alpha_i$.

If eccentricity ε is small—that is, when the bearing is lightly loaded—or when the supply pressure p_i is high, it is possible that cavitation would cease to exist (Dowson et al., 1985). In this case, there exists one set of α_c and α_s such that α_s equals $\alpha_c - 2\pi$ and Equation 1.31 holds. Since the boundary conditions described by Equations 1.31 and 1.33 still hold, the expression for the hydrodynamic fluid pressure given by Equation 1.36 remains valid. However, Equations 1.37 and 1.38 cannot be used any more to determine α_c and α_s since neither of the Equations 1.29 and 1.30 holds. Under this situation, the cyclic boundary condition $p(\alpha_c) = p(\alpha_c - 2\pi)$ applies. Equation 1.39 is derived by applying boundary condition $p(\alpha_c) = p(\alpha_c - 2\pi)$ on Equation A.8 (Wang and Khonsari, 2008).

$$\left(\omega - 2\dot{\phi}\right)\varepsilon\left(1-\varepsilon^2\right)^{0.5}\left(2\cos\alpha_c + \varepsilon\right) + 2\dot{\varepsilon}\left(2 + \varepsilon^2\right)\sin\alpha_c = 0 \qquad (1.39)$$

When cavitation does not exist, Equation 1.39 is used to determine α_c while $\pi < \alpha_c \le 2\pi + \alpha_i$. After determining α_c, $\alpha_s = \left(\alpha_c - 2\pi\right)$.

Thus, the hydrodynamic pressure profile in a long journal bearing can be predicted using Equation 1.36 with α_c, which can be determined using either Equation 1.37 or Equation 1.39 depending on whether the cavitation exists.

The hydrodynamic pressure given by Equation 1.36 can be normalized using Equation 1.40.

$$\bar{p} = \frac{C^2}{\left(\mu\omega R^2\right)p} \qquad (1.40)$$

1.3.2   Hydrodynamic Fluid Force of Long Journal Bearings

The hydrodynamic fluid force components in the radial (subscript ε) and tangential (subscript ϕ) directions of long journal bearings are given by Equation 1.41.

$$f_\varepsilon = RL\int_{\theta_s}^{\theta_c} p\cos\theta d\theta$$

$$ \qquad (1.41)$$

$$f_\phi = RL\int_{\theta_s}^{\theta_c} p\sin\theta d\theta$$

Integrating Equation 1.41 by parts and applying the boundary conditions (when cavitation exists, $p|_{\theta=\theta_c}=0$ and $p|_{\theta=\theta_s}=0$; when cavitation does not exist, $p|_{\theta=\theta_c}=p|_{\theta=\theta_s=\theta_c-2\pi}$) yields

$$f_\varepsilon = RL\left[p\sin\theta|_{\theta_s}^{\theta_c} - \int_{\theta_s}^{\theta_c}\frac{dp}{d\theta}\sin\theta d\theta\right] = -RL\int_{\theta_s}^{\theta_c}\frac{dp}{d\theta}\sin\theta d\theta$$

$$ \qquad (1.42)$$

$$f_\phi = RL\left[-p\cos\theta|_{\theta_s}^{\theta_c} + \int_{\theta_s}^{\theta_c}\frac{dp}{d\theta}\cos\theta d\theta\right] = RL\int_{\theta_s}^{\theta_c}\frac{dp}{d\theta}\cos\theta d\theta$$

Substituting Equation A.1 with C_1 given by Equation A.2 into Equation 1.42, solving the integrals (see Appendix B), and then substituting θ_c with α_c using Equation A.6, the hydrodynamic fluid force components in the radial and tangential directions are given by (Wang and Khonsari, 2008).

$$f_\varepsilon = -\frac{3\mu L R^3}{C^2}\left[\frac{(\omega-2\dot\phi)\varepsilon(\cos\alpha_c-\cos\alpha_s)^2}{(1-\varepsilon^2)(1-\varepsilon\cos\alpha_c)}\right.$$

$$\left.+2\dot\varepsilon\frac{(\alpha_c-\alpha_s+\sin\alpha_s\cos\alpha_s)(1-\varepsilon\cos\alpha_c)+(\varepsilon\cos^2\alpha_s-2\cos\alpha_s+\cos\alpha_c)\sin\alpha_c}{(1-\varepsilon^2)^{1.5}(1-\varepsilon\cos\alpha_c)}\right]$$

$$(1.43)$$

$$f_\phi = \frac{3\mu L R^3}{C^2}\left\{\frac{(\omega-2\dot\phi)\varepsilon}{(1-\varepsilon^2)^{1.5}(1-\varepsilon\cos\alpha_c)}[(1+2\varepsilon\cos\alpha_c)(\alpha_c-\alpha_s)\right.$$

$$-2(\varepsilon+\cos\alpha_c)(\sin\alpha_c-\sin\alpha_s)+(\sin\alpha_c\cos\alpha_c-\sin\alpha_s\cos\alpha_s)]$$

$$+\frac{2\dot\varepsilon}{(1-\varepsilon^2)^2(1-\varepsilon\cos\alpha_c)}[2\varepsilon(\cos\alpha_c-\varepsilon)-(2\varepsilon\cos\alpha_s+\sin^2\alpha_s)(1-\varepsilon\cos\alpha_c) \qquad (1.44)$$

$$\left.-\sin^2\alpha_c+3\varepsilon\sin\alpha_c(\alpha_c-\alpha_s)+(2\varepsilon^2+2-\varepsilon\cos\alpha_s)\sin\alpha_s\sin\alpha_c]\right\}$$

where "." represents d/dt.

In Equations 1.43 and 1.44, when cavitation exists, α_c is determined by Equation 1.37 and then α_s is determined by Equation 1.38; when cavitation does not exist, α_c is determined by Equation 1.39 and then $\alpha_s=(\alpha_c-2\pi)$.

Normalizing the time t by substituting $\bar t=\omega t$, the dimensionless radial and tangential components of the hydrodynamic fluid force in the journal bearing are defined as follows:

$$\bar f_\varepsilon = \frac{f_\varepsilon}{mC\omega^2} = -\frac{3\mu L R^3}{m\omega C^3}\left[\frac{(1-2\dot\phi)\varepsilon(\cos\alpha_c-\cos\alpha_s)^2}{(1-\varepsilon^2)(1-\varepsilon\cos\alpha_c)}\right.$$

$$\left.+2\dot\varepsilon\frac{(\alpha_c-\alpha_s+\sin\alpha_s\cos\alpha_s)(1-\varepsilon\cos\alpha_c)+(\varepsilon\cos^2\alpha_s-2\cos\alpha_s+\cos\alpha_c)\sin\alpha_c}{(1-\varepsilon^2)^{3/2}(1-\varepsilon\cos\alpha_c)}\right]$$

$$(1.45)$$

$$\bar f_\phi = \frac{f_\phi}{mC\omega^2} = \frac{3\mu L R^3}{m\omega C^3}\left\{\frac{(1-2\dot\phi)\varepsilon}{(1-\varepsilon^2)^{3/2}(1-\varepsilon\cos\alpha_c)}[(1+2\varepsilon\cos\alpha_c)(\alpha_c-\alpha_s)\right.$$

$$-2(\varepsilon+\cos\alpha_c)(\sin\alpha_c-\sin\alpha_s)+(\sin\alpha_c\cos\alpha_c-\sin\alpha_s\cos\alpha_s)]$$

$$+\frac{2\dot\varepsilon}{(1-\varepsilon^2)^2(1-\varepsilon\cos\alpha_c)}[2\varepsilon(\cos\alpha_c-\varepsilon)-(2\varepsilon\cos\alpha_s+\sin^2\alpha_s)(1-\varepsilon\cos\alpha_c)$$

$$\left.-\sin^2\alpha_c+3\varepsilon\sin\alpha_c(\alpha_c-\alpha_s)+(2\varepsilon^2+2-\varepsilon\cos\alpha_s)\sin\alpha_s\sin\alpha_c]\right\}$$

$$(1.46)$$

where "." represents $d/\omega dt$.

1.3.3   Static Performance of Long Journal Bearings

Similar to the short bearing analysis given in Section 1.2.3, assuming $\dot{\varepsilon} = \dot{\phi} = 0$, the fluid pressure, fluid force components, attitude angle, and the relation between eccentricity ratio ε and Sommerfeld number S are derived from Equations 1.36–1.39 and 1.43–1.44 and the force equilibrium between the fluid force $\mathbf{f}$ and the externally applied load $\mathbf{W}$ on the journal ($W = \sqrt{f_\varepsilon^2 + f_\phi^2}$ and $\tan\phi = -f_\phi/f_\varepsilon$). Table 1.2 compares the results derived under RFJ boundary condition with those based on the Reynolds boundary condition. The Reynolds boundary condition—sometimes referred to as the Swift-Stieber boundary condition— actually is one special case of RFJ boundary condition with some simplifications as ($\alpha_s = \alpha_i = 0$ and $p_i = 0$).

Based on the analysis in Section 1.3.2 and Table 1.2, the fluid film configuration, pressure distribution, and steady-state journal position in axially grooved journal bearing with different dimensionless fluid inlet pressures (in the range of $0 \le \bar{p}_i \le 1$) and different fluid inlet positions (in the range of $0 \le \Theta_i \le 90°$) are calculated and summarized below.

1.3.3.1   Comparison of Different Types of Boundary Conditions

Assuming an inlet condition of ($\Theta_i = 0$, $\bar{p}_i = 0$), Figure 1.5 compares how the eccentricity ratio (ε) varies as a function of the Sommerfeld number (S) based on the RFJ boundary condition and the half-Sommerfeld boundary condition. Figure 1.6 shows how the attitude angle (ϕ) varies as a function of eccentricity ratio (ε) based on different boundary conditions. Since the journal position can be fully defined as (ε, ϕ) in a polar coordinate, Figure 1.6 also compares the journal position as a function of eccentricity ratio (ε) based on different boundary conditions.

Figure 1.5 shows that the eccentricity ratio ε predicted based on the assumed half-Sommerfeld boundary condition is always overestimated compared to that predicted based on the RFJ boundary condition. However, Figure 1.6 shows that the attitude angle (ϕ) predicted based on the half-Sommerfeld boundary condition is underestimated when ε is less than about 0.85. These inaccuracies are caused by neglecting the cavitation developed in the geometrically diverging region under the assumption of half-Sommerfeld boundary condition. The detailed cavitation will be shown together with the fluid-film configuration and pressure distribution presented in the next subsection.

1.3.3.2   Influence of Fluid Inlet Pressure

Assuming the fluid inlet position held at $\Theta_i = 0°$, the influence of different fluid inlet pressures ($\bar{p}_i = 0$, 0.5, or 1.0) on the fluid film configuration, pressure

Table 1.2 Static performances based on different types of boundary condition

RFJ boundary condition	Reynolds boundary condition
Pressure solution domain: $\alpha_s \leq \alpha \leq \alpha_c$,	Pressure solution domain: $0 \leq \alpha \leq \alpha_c$,

RFJ boundary condition:

$p = p_i + 3\mu\omega R^2 \varepsilon / \left[C^2 \left(1-\varepsilon^2\right)^{1.5} \left(1-\varepsilon\cos\alpha_c\right) \right]$

$\left[2\left(\sin\alpha - \sin\alpha_i\right)\left(1+\varepsilon\cos\alpha_c\right) + \left(2\cos\alpha_c + \varepsilon\right) \right.$;

$\left. \left(\alpha_i - \alpha\right) - \varepsilon\left(\sin\alpha\cos\alpha - \sin\alpha_i\cos\alpha_i\right) \right]$
elsewhere, $p = 0$.

Reynolds boundary condition:

$$p = \frac{6\mu\omega R^2}{C^2\left(1-\varepsilon^2\right)^{1.5}} \left[\alpha - \varepsilon\sin\alpha \right.$$

$$\left. - \frac{\left(\alpha + \dfrac{\varepsilon^2}{2}\alpha - 2\varepsilon\sin\alpha + \dfrac{\varepsilon^2\sin 2\alpha}{4}\right)}{\left(1-\varepsilon\cos\alpha_c\right)} \right] ;$$

elsewhere, $p = 0$.

If cavitation exits, α_c and α_s are obtained from

$\varepsilon\left[\left(2+\varepsilon\cos\alpha_c\right)\left(\sin\alpha_c - \sin\alpha_i\right) - \varepsilon\sin\alpha_i\right.$

$\left.\left(\cos\alpha_c - \cos\alpha_i\right) + \left(2\cos\alpha_c + \varepsilon\right)\left(\alpha_i - \alpha_c\right)\right]$

$+ C^2 p_i \left(1-\varepsilon^2\right)^{1.5}\left(1-\varepsilon\cos\alpha_c\right)/\left(3\mu\omega R^2\right) = 0$

$\varepsilon\left[2\left(\sin\alpha_s - \sin\alpha_i\right)\left(1+\varepsilon\cos\alpha_c\right) + \left(2\cos\alpha_c + \varepsilon\right)\right.$

$\left.\left(\alpha_i - \alpha_s\right) - \varepsilon\left(\sin\alpha_s\cos\alpha_s - \sin\alpha_i\cos\alpha_i\right)\right]$

$+ C^2 p_i \left(1-\varepsilon^2\right)^{1.5}\left(1-\varepsilon\cos\alpha_c\right)/\left(3\mu\omega R^2\right) = 0$

If no cavitation exits, α_c and α_s are obtained from

$\alpha_c = \cos^{-1}\left(-\varepsilon/2\right), \alpha_s = \alpha_c - 2\pi$

α_c is obtained from

$\varepsilon\left(\alpha_c - \sin\alpha_c\cos\alpha_c\right)$

$= 2\left(\sin\alpha_c - \alpha_c\cos\alpha_c\right)$

Fluid force components:

$$f_\varepsilon = -\frac{3\mu\omega LR^3}{C^2}\frac{\varepsilon\left(\cos\alpha_c - \cos\alpha_s\right)^2}{\left(1-\varepsilon^2\right)\left(1-\varepsilon\cos\alpha_c\right)}$$

$$f_\phi = \frac{3\mu\omega LR^3}{C^2}\frac{\varepsilon}{\left(1-\varepsilon^2\right)^{1.5}\left(1-\varepsilon\cos\alpha_c\right)}$$

$\left[\left(1+2\varepsilon\cos\alpha_c\right)\left(\alpha_c - \alpha_s\right) - 2\left(\varepsilon + \cos\alpha_c\right)\right.$

$\left.\left(\sin\alpha_c - \sin\alpha_s\right) + \left(\sin\alpha_c\cos\alpha_c - \sin\alpha_s\cos\alpha_s\right)\right]$

Fluid force components:

$$f_\varepsilon = -\frac{3\mu\omega LR^3}{C^2}\frac{\varepsilon\left(1-\cos\alpha_c\right)^2}{\left(1-\varepsilon^2\right)\left(1-\varepsilon\cos\alpha_c\right)}$$

$$f_\phi = \frac{6\mu\omega LR^3}{C^2}\frac{\sin\alpha_c - \alpha_c\cos\alpha_c}{\left(1-\varepsilon^2\right)^{0.5}\left(1-\varepsilon\cos\alpha_c\right)}$$

Relation between Sommerfeld number and eccentricity ratio:

$$S = \frac{\left(1-\varepsilon^2\right)^{1.5}\left(1-\varepsilon\cos\alpha_c\right)}{3\pi\varepsilon\sqrt{\left(1-\varepsilon^2\right)\left(\cos\alpha_c - \cos\alpha_s\right)^4 + A_1^2}}$$

Relation between Sommerfeld number and eccentricity ratio:

$$S = \frac{\left(1-\varepsilon^2\right)\left(1-\varepsilon\cos\alpha_c\right)}{3\pi\varepsilon\sqrt{\left(1-\cos\alpha_c\right)^4 + \left(1-\varepsilon^2\right)A_1^2}}$$

Attitude angle:

$\phi = \tan^{-1}\left[A_1\left(1-\varepsilon^2\right)^{-0.5}\left(\cos\alpha_c - \cos\alpha_s\right)^{-2}\right]$
where $A_1 = \left(1+2\varepsilon\cos\alpha_c\right)\left(\alpha_c - \alpha_s\right)$
$- 2\left(\varepsilon + \cos\alpha_c\right)$
$\left(\sin\alpha_c - \sin\alpha_s\right) + \sin\alpha_c\cos\alpha_c - \sin\alpha_s\cos\alpha_s$.

Attitude angle:

$\phi = \tan^{-1}\left[A_1\left(1-\varepsilon^2\right)^{0.5}\left(1-\cos\alpha_c\right)^{-2}\right]$
where $A_1 = \alpha_c - \sin\alpha_c\cos\alpha_c$.

From Wang and Khonsari (2008) © Elsevier Limited.

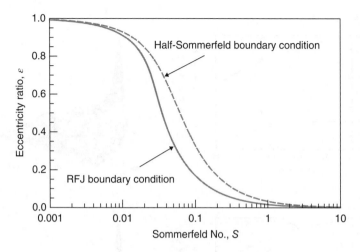

Figure 1.5   Comparison of the curve S vs. ε based on different boundary conditions ($\Theta_i = 0$, $\bar{p}_i = 0$). From Wang and Khonsari (2008) © Elsevier Limited.

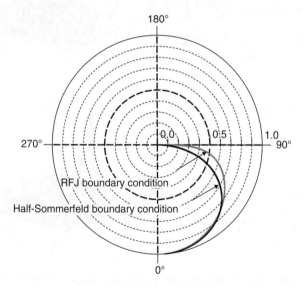

Figure 1.6   Comparison of the journal position based on different boundary conditions ($\Theta_i = 0$, $\bar{p}_i = 0$). From Wang and Khonsari (2008) © Elsevier Limited.

distribution, and steady state equilibrium position ($C\varepsilon$, ϕ) are summarized and discussed in this subsection.

Figure 1.7 shows how the fluid film configuration and the pressure distribution vary as a function of the eccentricity ratio ε in terms of the absolute circumferential

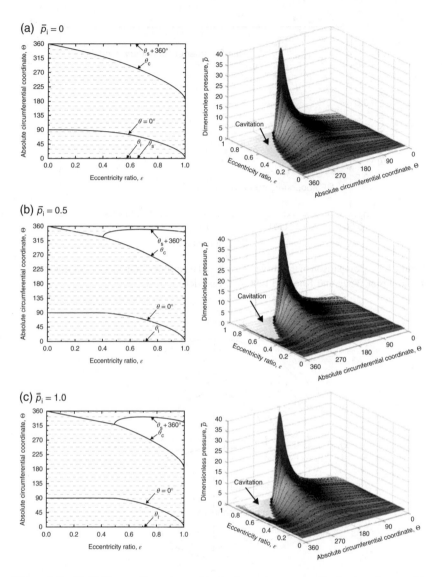

Figure 1.7   Fluid film configuration and hydrodynamic pressure distribution change with changing the steady-state eccentricity ratio ε in terms of the absolute circumferential coordinate Θ ($\Theta_i = 0$) at different supply pressures $\bar{p}_i = 0$, 0.5, and 1.0. From Wang and Khonsari (2008) © Elsevier Limited.

coordinate Θ, which starts from the fixed position at the upper load line. The position of $\theta = 0$ is defined at $\Theta = \phi$ since $\Theta = \theta + \phi$.

With the fluid inlet condition of ($\Theta_i = 0$, $\bar{p}_i = 0$), Figure 1.7a shows that the fluid pressure starting position θ_s coincides with the fluid inlet position θ_i. The fluid pressure always starts from the fluid inlet position. It also shows that cavitation always exists with the fluid inlet condition of ($\Theta_i = 0$, $\bar{p}_i = 0$). The cavitation region shrinks as the steady state eccentricity ratio ε decreases.

Figure 1.7b shows that cavitation exists only if the steady state eccentricity ratio ε is greater than about 0.4 with the fluid inlet condition of ($\Theta_i = 0$, $\bar{p}_i = 0.5$). When $\varepsilon > 0.4$, the cavitation region shrinks to a smaller portion of the journal surface with a smaller ε down to 0.4; otherwise, a full 2π fluid film exists in the fluid-film journal bearings.

Figure 1.7c shows that cavitation exists only if the steady-state eccentricity ratio ε is greater than about 0.49 with the fluid inlet condition of ($\Theta_i = 0$, $\bar{p}_i = 1.0$). When $\varepsilon > 0.49$, the cavitation region shrinks to a smaller portion of the journal surface with a smaller ε down to 0.49; otherwise, a full 2π fluid film exists in the fluid-film journal bearings.

Figure 1.8 compares the pressure distributions corresponding to different fluid inlet pressures $\bar{p}_i = 0$, $\bar{p}_i = 0.5$, and $\bar{p}_i = 1.0$, respectively with $\varepsilon = 0.5$. Cavitation is shown to exist for all of the three fluid inlet conditions. However, the cavitation region shrinks as the inlet pressure $\bar{p}_i$ increases from 0 to 1.0. The peak pressure increases with increasing the fluid inlet pressure $\bar{p}_i$ from 0 to 1.0.

Figure 1.9 shows a comparison of the curve S vs. ε corresponding to different fluid inlet pressure $\bar{p}_i = 0$, $\bar{p}_i = 0.5$, and $\bar{p}_i = 1.0$, respectively. The effect of fluid inlet pressure on the curve S vs. ε is subtle. For a given Sommerfeld number S, the eccentricity ratio ε decreases slightly with increasing the fluid inlet pressure $\bar{p}_i$ from 0 to 1.0.

A comparison of the steady-state equilibrium positions corresponding to different fluid inlet pressures $\bar{p}_i = 0$, $\bar{p}_i = 0.5$, and $\bar{p}_i = 1.0$ is shown in Figure 1.10. It is

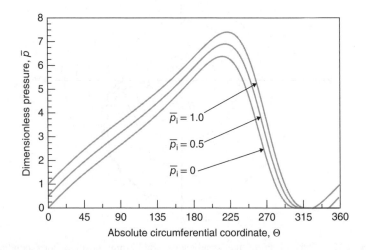

Figure 1.8   Comparison of the pressure distributions with $\varepsilon = 0.5$ and $\Theta_i = 0$. From Wang and Khonsari (2008) © Elsevier Limited.

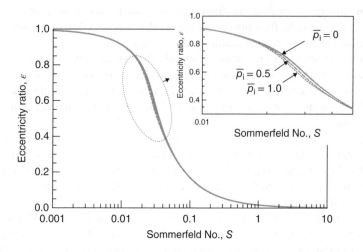

Figure 1.9 Comparison of the curve S vs. ε corresponding to different fluid inlet pressures $(\Theta_i = 0)$. From Wang and Khonsari (2008) © Elsevier Limited.

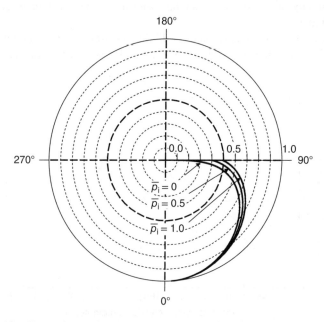

Figure 1.10 Comparison of the steady-state equilibrium position corresponding to different fluid inlet pressures $(\Theta_i = 0)$. From Wang and Khonsari (2008) © Elsevier Limited.

shown that the effect of fluid inlet pressure on the steady-state equilibrium position is substantial. For a given eccentricity ratio ε, the attitude angle ϕ increases as the fluid inlet pressure $\bar{p}_i$ increases from 0 to 1.0. In addition, under the fluid inlet condition of ($\Theta_i = 0$, $\bar{p}_i = 0.5$), the attitude angle ϕ stays at 90° when the eccentricity ratio ε is in the range of 0–0.4. In other words, irrespective of how the eccentricity changes within the range of 0–0.4, since the radial hydrodynamic fluid force component f_ε does not exist, the journal is always in a force equilibrium and is free to be moved to and then stay at any position with a new eccentricity and the same attitude angle. This feature makes the rotor bearing system very sensitive to external perturbation in the horizontal direction. Any external perturbation in the horizontal direction will move the journal to a new position and no force is available to bring it back to its original position. This is sometimes referred as inherent instability to any external perturbations in the horizontal direction, which often exists in real applications. With a higher inlet pressure ($\Theta_i = 0$, $\bar{p}_i = 1.0$), the range where the attitude angle ϕ stays at 90° regardless of any change in eccentricity ratio ε expands to 0–0.49.

1.3.3.3 Influence of Fluid Inlet Position

We will examine the influence of different fluid inlet positions ($\Theta_i = 0°$, 45°, or 90°) on the fluid film configuration, fluid pressure distribution, and steady-state journal position ($C\varepsilon$, ϕ) in this subsection. The fluid inlet pressure is held at $\bar{p}_i = 0$.

Figure 1.11 shows how the fluid film configuration and pressure distribution vary as a function of the eccentricity ratio ε corresponding to different fluid inlet conditions ($\Theta_i = 45°$, $\bar{p}_i = 0$) and ($\Theta_i = 90°$, $\bar{p}_i = 0$), respectively. Figures 1.7a and 1.11a and b show that cavitation always exists with fluid inlet pressure $\bar{p}_i = 0$ and the cavitation region extends to a larger portion of the journal surface as the fluid inlet position angle Θ_i increases from 0° to 90°.

Assuming the eccentricity ratio $\varepsilon = 0.5$, Figure 1.12 shows a comparison of the fluid pressure distributions corresponding to different fluid inlet position $\Theta_i = 0$, $\Theta_i = 45°$, and $\Theta_i = 90°$, respectively. Cavitation region extends to a larger portion of the journal surface as the fluid inlet position Θ_i changes from 0° to 90°. With increasing the fluid inlet position Θ_i from 0° to 90°, the peak pressure decreases and moves to a higher angle position.

With the fluid inlet pressure $\bar{p}_i = 0$, Figures 1.13 and 1.14 show a comparison of the curve S vs. ε and a comparison of the steady-state equilibrium position corresponding to three different fluid inlet positions $\Theta_i = 0$, $\Theta_i = 45°$, and $\Theta_i = 90°$, respectively.

Figure 1.13 shows that for a given Sommerfeld number S, the eccentricity ratio ε increases with increasing the fluid inlet position Θ_i from 0° to 90°. Figure 1.14 shows that, for a given eccentricity ratio ε, the attitude angle ϕ decreases with increasing the fluid inlet position Θ_i from 0° to 90°. For lightly loaded journal bearings (that is small steady-state eccentricity ratio ε), the fluid inlet position effect is more pronounced.

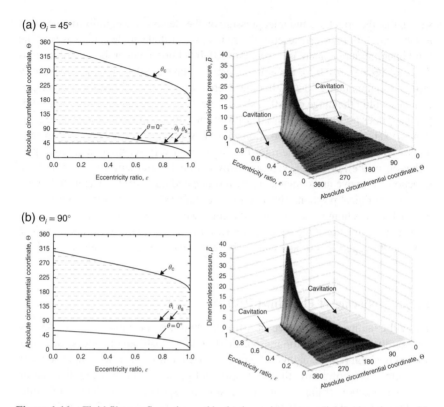

Figure 1.11   Fluid film configuration and hydrodynamic pressure distribution change with changing the steady-state eccentricity ratio ε in terms of the absolute circumferential coordinate Θ ($\bar{p}_i = 0$) for different inlet positions $\Theta_i = 45°$ and $90°$. From Wang and Khonsari (2008) © Elsevier Limited.

Based on the results presented in Figures 1.7, 1.8, 1.9, 1.10, 1.11, 1.12, 1.13, and 1.14, a set of curve-fitting functions are obtained and presented in Appendix C. Under different fluid supply conditions, Equations C.1–C.20 define how θ_c (circumferential location where the fluid film ruptures), θ_s (fluid pressure starting position), and ϕ (attitude angle) change with changing the steady-state eccentricity ratio ε and how the steady-state eccentricity ratio ε changes as the Sommerfeld number (S) changes.

1.4   Finite Bearing Solution

As discussed in the previous sections, for finite bearings with length-over-diameter ratio more than 0.5 while not exceeding 2, either the infinitely short bearing theory or the infinitely long bearing theory can only provide a rough estimation of the bearing performance. If more accurate prediction of the bearing performance is expected, numerical solutions to the Reynolds equations would be necessary.

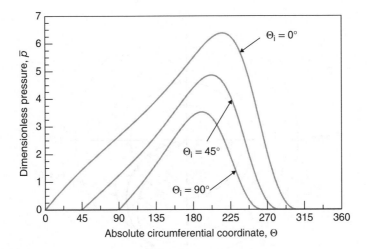

Figure 1.12 Comparison of the pressure distributions with $\varepsilon = 0.5$ and $\bar{p}_i = 0$. From Wang and Khonsari (2008) © Elsevier Limited.

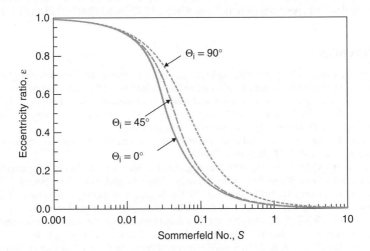

Figure 1.13 Comparison of the curve S vs. ε corresponding to different fluid inlet positions ($\Theta_i = 0$). From Wang and Khonsari (2008) © Elsevier Limited.

Two methods commonly used to obtain the numerical solutions to the Reynolds equations are finite element methods (FEA) and finite difference methods (FDM). Application of FEA and/or FDM to numerically solve the full Reynolds equations is out of the scope of this book. Detailed applications of numerical methods have been presented by Khonsari and Booser (2008), Khonsari and Wang (1991).

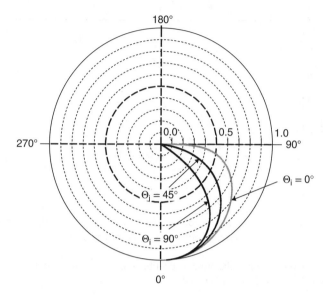

Figure 1.14 Comparison of the steady-state equilibrium position corresponding to different fluid inlet positions ($\bar{p}_i = 0$). From Wang and Khonsari (2008) © Elsevier Limited.

References

Dowson, D., Ruddy, A.V., Sharp, R.S., Taylor, C.M., 1985, "Analysis of the Circumferentially Grooved Journal Bearing with Consideration of Lubricant Film Reformation," *Proceedings of the Institution of Mechanical Engineers, Part C: Journal of Mechanical Engineering Science*, **199**, pp. 27–34.

Floberg, L. 1957, "The Infinite Journal Bearing, Considering Vaporization," Transactions of Chalmers University of Technology, 189, Goteborg, Sweden, pp. 1–82.

Hashimoto, H., Wada, S., 1982, "Influence of Inertia Forces on Stability of Turbulent Journal Bearings," *Bulletin of the JSME*, **25** (202), pp. 653–662.

Hashimoto, H., Wada, S., Ito, J.I., 1987, "An Application of Short Bearing Theory to Dynamic Characteristic Problems of Turbulent Journal Bearings," *ASME Journal of Tribology*, **109**, pp. 307–314.

Jakobsson, B., Floberg, L., 1957, "The Finite Journal Bearing, Considering Vaporization," *Transactions of Chalmers University of Technology*, **190**, Goteborg, Sweden, pp. 1–116.

Khonsari, M.M., Wang, S.H., 1991, "On the Fluid-Solid Interaction in Reference to Thermohydrodynamic Analysis of Journal Bearings," *ASME Journal of Tribology*, **113**, pp. 398–404.

Khonsari, M.M., Booser, E.R., 2008, *Applied Tribology: Bearing Design and Lubrication*, 2nd edition, John Wiley & Sons, Inc., New York.

Lund, J.W., 1966, "Self-excited, Stationary Whirl Orbits of a Journal in a Sleeve Bearings," PhD thesis, Department of Mechanics, Rensselaer Polytechnic Institute, Troy.

Ng, C.W., Pan, C.H.T., 1965, "A Linearized Turbulent Lubrication Theory," *ASME Journal of Basic Engineering*, **87**, pp. 675–688.

Wang, J.K., Khonsari, M.M., 2006, "Application of Hopf Bifurcation Theory to the Rotor-bearing System with Turbulent Effects," *Tribology International*, **39**(7), pp. 701–714.

Wang, J.K., Khonsari, M.M., 2008, "Effects of Oil Inlet Pressure and Inlet Position of Axially Grooved Infinitely Long Journal Bearings, Part I: Analytical Solutions and Static Performance," *Tribology International*, **41**, pp. 119–131.

Zhang, Y., 1989, "Dynamic Properties of Flexible Journal Bearings of Infinite Width Considering Oil Supply Position and Pressure," *Wear*, **130**, pp. 53–68.

2

Governing Equations for Dynamic Analysis

2.1 Equation of Motion

Consider a rotor bearing system consisting of a rigid, perfectly balanced, and symmetrical rotor (with mass 2m) horizontally supported on two identical fluid-film journal bearings as shown in Figure 2.1a. Due to the symmetry of this rotor bearing system, only one of the fluid-film journal bearings needs to be analyzed. Figure 2.1b depicts the orbit of the rotor journal as it travels within the bearing clearance circle. The definition of the bearing clearance circle and its implications have been introduced in Chapter 1.

In Figure 2.1, O_b represents the center of the journal bearing; O_{js} denotes the steady state equilibrium position of the journal center; O_{jd} is the dynamic position of the journal center; W is the load per bearing; and ω is the rotor rotating speed. ε and ϕ are dynamic eccentricity ratio and attitude angle, respectively. These two parameters can vary with time t. C is the radial clearance (i.e., $C = R_b - R_j$, R_b is the inside radius of bearing bushing, R_j is the radius of rotor journal); f_ε and f_ϕ are the radial and tangential components of the hydrodynamic fluid force applied on the rotor journal; and x and y are the Cartesian coordinates in the horizontal and vertical directions, respectively.

Thermohydrodynamic Instability in Fluid-Film Bearings, First Edition.
J. K. Wang and M. M. Khonsari.
© 2016 John Wiley & Sons, Ltd. Published 2016 by John Wiley & Sons, Ltd.

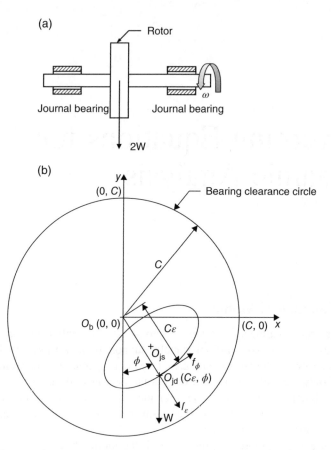

Figure 2.1   (a) Model of a rigid rotor supported by two identical journal bearings. (b) Sketch of the journal orbit within the bearing clearance circle (Wang and Khonsari, 2006)

For this rotor-bearing system, its equations of motion derived from Newton's second law $\sum F = ma$ in polar coordinates (ε, ϕ) are (Wang and Khonsari, 2006)

$$\varepsilon\text{-direction}: mC\ddot{\varepsilon} - mC\varepsilon\dot{\phi}^2 = f_\varepsilon + W\cos\phi \tag{2.1}$$

$$\phi\text{-direction}: mC\varepsilon\ddot{\phi} + 2mC\dot{\varepsilon}\dot{\phi} = f_\phi - W\sin\phi \tag{2.2}$$

In Equations 2.1 and 2.2, "." represents d/dt and m is the rotor mass per bearing. The hydrodynamic force components f_ε and f_ϕ will be defined in the next two sections accordingly.

Normalizing the time t parameter by substituting $\bar{t} = \omega t$, the dimensionless forms of the equations of motion are given as follows (Wang and Khonsari, 2006):

$$\ddot{\varepsilon} - \varepsilon\dot{\phi}^2 - \bar{f}_\varepsilon - \frac{W}{mC\omega^2}\cos\phi = 0 \tag{2.3}$$

$$\ddot{\phi} + \frac{2\dot{\varepsilon}\dot{\phi}}{\varepsilon} - \frac{\bar{f}_\phi}{\varepsilon} + \frac{W}{mC\omega^2\varepsilon}\sin\phi = 0 \qquad (2.4)$$

where "." represents $d/\omega dt$. $\bar{f}_\varepsilon$ and $\bar{f}_\phi$ are the dimensionless radial and tangential components of the hydrodynamic fluid force in the journal bearing.

At steady-state position $\dot{\varepsilon} = \dot{\phi} = 0$, Equations 2.1 and 2.2 degenerate to $f_{\varepsilon_s} = -W\cos\phi_s$ and $f_{\phi_s} = W\sin\phi_s$. As Figure 2.1b shows, these two simple equations have very obvious physical meanings. The steady-state fluid force balances the static weight assigned to the fluid film bearing.

2.2 Decomposition of the Equations of Motion Based on Short Bearing Theory

In Section 1.2, we derived the radial and tangential components of the hydrodynamic fluid force exerted on the rotor journal by the fluid film short journal bearing. Assuming constant fluid viscosity, for convenience, the expressions of the dimensionless radial and tangential components of the hydrodynamic fluid force are repeated as follows:

$$\bar{f}_\varepsilon = \frac{f_\varepsilon}{mC\omega^2} = -\frac{\mu RL^3}{2mC^3\omega}\left\{ (1-2\dot{\phi})\left[\frac{2(a_z-2b_z)\varepsilon^2+2b_z}{(1-\varepsilon^2)^2} - \frac{b_z}{\varepsilon}\ln\left(\frac{1+\varepsilon}{1-\varepsilon}\right) \right] \right.$$
$$\left. + \pi\dot{\varepsilon}\left[\frac{a_z+5b_z+2(a_z-3b_z)\varepsilon^2-\frac{2b_z}{\varepsilon^2}}{(1-\varepsilon^2)^{5/2}} + \frac{2b_z}{\varepsilon^2} \right] \right\} \qquad (2.5)$$

$$\bar{f}_\phi = \frac{f_\phi}{mC\omega^2} = \frac{\mu RL^3}{2mC^3\omega\varepsilon}\left\{ \pi(1-2\dot{\phi})\left[\frac{2b_z+(a_z-3b_z)\varepsilon^2}{2(1-\varepsilon^2)^{3/2}} - b_z \right] \right.$$
$$\left. + 4\dot{\varepsilon}\left[\frac{b_z+(a_z-2b_z)\varepsilon^2}{(1-\varepsilon^2)^2} - \frac{b_z}{2\varepsilon}\ln\left(\frac{1+\varepsilon}{1-\varepsilon}\right) \right] \right\} \qquad (2.6)$$

To solve the nonlinear equations of motion (Eqs. 2.3 and 2.4 with $\bar{f}_\varepsilon$ and $\bar{f}_\phi$ given in Eqs. 2.5 and 2.6), these two second-order nonlinear differential equations are decomposed into four first-order equations.

Letting $x_1 = \varepsilon$, $x_2 = \dot{\varepsilon}$, $x_3 = \phi$, $x_4 = \dot{\phi}$, $\Gamma = \mu RL^3/(2mC^{2.5}g^{0.5})$ and $\bar{\omega} = \omega\sqrt{C/g}$, we arrive at the following equations (Wang and Khonsari, 2006):

$$\dot{x}_1 = x_2 \qquad (2.7)$$

$$
\begin{aligned}
\dot{x}_2 =& x_1 x_4^2 + \frac{f_\varepsilon}{mC\omega^2} + \frac{g}{C\omega^2}\cos x_3 \\
=& x_1 x_4^2 - \frac{\Gamma}{\bar{\omega}}\left\{(1-2x_4)\left[\frac{2(a_z-2b_z)x_1^2+2b_z}{\left(1-x_1^2\right)^2} - \frac{b_z}{x_1}\ln\left(\frac{1+x_1}{1-x_1}\right)\right]\right. \\
& \left. + \pi x_2 \left[\frac{a_z + 5b_z + 2(a_z-3b_z)x_1^2 - \left(2b_z/x_1^2\right)}{\left(1-x_1^2\right)^{5/2}} + \frac{2b_z}{x_1^2}\right]\right\} + \frac{1}{\bar{\omega}^2}\cos x_3
\end{aligned}
\tag{2.8}
$$

$$
\dot{x}_3 = x_4
\tag{2.9}
$$

$$
\begin{aligned}
\dot{x}_4 =& -\frac{2x_2 x_4}{x_1} + \frac{f_\phi}{x_1 mC\omega^2} - \frac{g}{C\omega^2 x_1}\sin x_3 \\
=& -\frac{2x_2 x_4}{x_1} + \frac{\Gamma}{\bar{\omega}x_1^2}\left\{\pi(1-2x_4)\left[\frac{2b_z + (a_z-3b_z)x_1^2}{2\left(1-x_1^2\right)^{3/2}} - b_z\right]\right. \\
& \left. + 4x_2\left[\frac{b_z + (a_z-2b_z)x_1^2}{\left(1-x_1^2\right)^2} - \frac{b_z}{2x_1}\ln\left(\frac{1+x_1}{1-x_1}\right)\right]\right\} - \frac{1}{\bar{\omega}^2 x_1}\sin x_3
\end{aligned}
\tag{2.10}
$$

where g is the gravitational constant. Γ is a dimensionless bearing's characteristic constant assuming that the fluid viscosity μ is constant.

The above system of Equations 2.7–2.10 is of the form

$$
\dot{\mathbf{x}} = \mathbf{f}(\mathbf{x}, \bar{\omega})
\tag{2.11}
$$

The steady-state equilibrium position $\mathbf{x}_s$ in terms of $(x_{1s} = \varepsilon_s,\ x_{2s} = \dot{\varepsilon}_s,\ x_{3s} = \phi_s,$ $x_{4s} = \dot{\phi}_s)$ can be found analytically by letting $\mathbf{f}(\mathbf{x}_s, \bar{\omega}) = 0$. The subscript s here denotes the steady-state equilibrium position.

For short bearings, the resulting equations for the steady state equilibium position of the system with consideration of the turbulent effects are given in Equation 2.12 (Wang and Khonsari, 2006). Since $\dot{\varepsilon}_s = 0$ and $\dot{\phi}_s = 0$, the steady-state equilibrium position can be rewriten as $\mathbf{x}_s = (\varepsilon_s, 0, \phi_s, 0)$.

$$
\left[\frac{2(a_z-2b_z)x_{1s}^2 + 2b_z}{\left(1-x_{1s}^2\right)^2} - \frac{b_z}{x_{1s}}\ln\left(\frac{1+x_{1s}}{1-x_{1s}}\right)\right]^2 + \frac{\pi^2}{x_{1s}^2}\left[\frac{2b_z + (a_z-3b_z)x_{1s}^2}{2\left(1-x_{1s}^2\right)^{3/2}} - b_z\right]^2 = \left(\frac{1}{\Gamma\bar{\omega}}\right)^2
$$

$$
x_{2s} = 0
$$

$$
x_{3s} = \tan^{-1}\left(\frac{\frac{\pi}{2}\left[2b_z + (a_z-3b_z)x_{1s}^2\right]\left(1-x_{1s}^2\right)^{1/2} - \pi b_z\left(1-x_{1s}^2\right)^2}{\left[2(a_z-2b_z)x_{1s}^2 + 2b_z\right]x_{1s} - b_z\left(1-x_{1s}^2\right)^2\ln\left(\frac{1+x_{1s}}{1-x_{1s}}\right)}\right)
$$

$$
x_{4s} = 0
$$

$$
\tag{2.12}
$$

Since $\Gamma\bar{\omega} = 2S\pi(L/D)^2$, the first equation in Equations 2.12 shows a relationship between the Sommerfeld number S and the eccentricity, ε_s, and the third equation gives an expression for the attitude angle, ϕ_s, as a function of Sommerfeld number.

2.2.1 Laminar Flow Simplification

The above equations are valid for short bearing with provision for turbulent flow. Under laminar flow condition, $a_z = 1$ and $b_z = 0$. Therefore, the radial and tangential components of the hydrodynamic fluid force (Eqs. 2.5 and 2.6) are simplified as (Wang and Khonsari, 2006)

$$f_\varepsilon = -\frac{\mu R L^3}{2C^2}\left[\frac{2(\omega - 2\dot{\phi})\varepsilon^2}{(1-\varepsilon^2)^2} + \frac{\pi\dot{\varepsilon}(1+2\varepsilon^2)}{(1-\varepsilon^2)^{5/2}}\right] \tag{2.13}$$

$$f_\phi = \frac{\mu R L^3}{2C^2}\left[\frac{\pi(\omega - 2\dot{\phi})\varepsilon}{2(1-\varepsilon^2)^{3/2}} + \frac{4\varepsilon\dot{\varepsilon}}{(1-\varepsilon^2)^2}\right] \tag{2.14}$$

The resulting decomposed equation of motion (Eqs. 2.7–2.10) can be simplified as (Wang and Khonsari, 2006)

$$\dot{x}_1 = x_2 \tag{2.15}$$

$$\dot{x}_2 = x_1 x_4^2 - \frac{\Gamma}{\bar{\omega}}\left[\frac{2x_1^2(1-2x_4)}{(1-x_1^2)^2} + \frac{\pi(1+2x_1^2)x_2}{(1-x_1^2)^{5/2}}\right] + \frac{1}{\bar{\omega}^2}\cos x_3 \tag{2.16}$$

$$\dot{x}_3 = x_4 \tag{2.17}$$

$$\dot{x}_4 = -\frac{2x_2 x_4}{x_1} + \frac{\Gamma}{\bar{\omega}}\left[\frac{\pi(1-2x_4)}{2(1-x_1^2)^{3/2}} + \frac{4x_2}{(1-x_1^2)^2}\right] - \frac{1}{\bar{\omega}^2 x_1}\sin x_3 \tag{2.18}$$

The resulting Equations 2.12 for the steady state equilibrium position of the system with provision of laminar flow can be simplified as (Wang and Khonsari, 2006)

$$\frac{x_{1s}\sqrt{16x_{1s}^2 + \pi^2(1-x_{1s}^2)}}{(1-x_{1s}^2)^2} = \frac{1}{S\pi\left(\dfrac{L}{D}\right)^2}$$

$$x_{2s} = 0 \tag{2.19}$$

$$x_{3s} = \tan^{-1}\left(\frac{\pi\sqrt{1-x_{1s}^2}}{4x_{1s}}\right)$$

$$x_{4s} = 0$$

These expressions are consistent to those obtained by direct solution of the Reynolds equation based on the infinitely short bearing assumption and laminar flow condition in Chapter 1.

2.3 Decomposition of the Equations of Motion Based on Long Bearing Theory

Similar analysis can be performed for long bearings by referring to the results of Section 1.3. For convenience, the long bearing expressions of the dimensionless radial and tangential components of the hydrodynamic fluid force are repeated below (Wang and Khonsari, 2008):

$$\bar{f}_\varepsilon = \frac{f_\varepsilon}{mC\omega^2} = -\frac{3\mu LR^3}{m\omega C^3}\left[\frac{(1-2\dot\phi)\varepsilon(\cos\alpha_c-\cos\alpha_s)^2}{(1-\varepsilon^2)(1-\varepsilon\cos\alpha_c)}\right.$$

$$\left.+2\dot\varepsilon\frac{(\alpha_c-\alpha_s+\sin\alpha_s\cos\alpha_s)(1-\varepsilon\cos\alpha_c)+(\varepsilon\cos^2\alpha_s-2\cos\alpha_s+\cos\alpha_c)\sin\alpha_c}{(1-\varepsilon^2)^{3/2}(1-\varepsilon\cos\alpha_c)}\right] \tag{2.20}$$

$$\bar{f}_\phi = \frac{f_\phi}{mC\omega^2} = \frac{3\mu LR^3}{m\omega C^3}\left\{\frac{(1-2\dot\phi)\varepsilon}{(1-\varepsilon^2)^{3/2}(1-\varepsilon\cos\alpha_c)}[(1+2\varepsilon\cos\alpha_c)(\alpha_c-\alpha_s)\right.$$

$$-2(\varepsilon+\cos\alpha_c)(\sin\alpha_c-\sin\alpha_s)+(\sin\alpha_c\cos\alpha_c-\sin\alpha_s\cos\alpha_s)]$$

$$+\frac{2\dot\varepsilon}{(1-\varepsilon^2)^2(1-\varepsilon\cos\alpha_c)}[2\varepsilon(\cos\alpha_c-\varepsilon)-(2\varepsilon\cos\alpha_s+\sin^2\alpha_s)(1-\varepsilon\cos\alpha_c) \tag{2.21}$$

$$\left.-\sin^2\alpha_c+3\varepsilon\sin\alpha_c(\alpha_c-\alpha_s)+(2\varepsilon^2+2-\varepsilon\cos\alpha_s)\sin\alpha_s\sin\alpha_c]\right\}$$

where "." represents $d/\omega dt$.

In Equations 2.20 and 2.21, α_s corresponds to θ_s, which defines fluid pressure starting position and α_c corresponds to θ_c, the starting position of fluid cavitation. The relation between α_s and θ_s and the relation between α_c and θ_c are described by Equations A.5 and A.6, respectively. α_s and α_c are all a function of ε. α_i corresponds to θ_i, which defines the fluid inlet position. The relation between α_i and θ_i is given by Equation A.7. α_i is defined by Equation 2.22 (Wang and Khonsari, 2008).

$$\cos\alpha_i = \frac{\varepsilon+\cos\theta_i}{1+\varepsilon\cos\theta_i} = \frac{\varepsilon+\cos(\Theta_i-\phi)}{1+\varepsilon\cos(\Theta_i-\phi)} \tag{2.22}$$

where Θ_i is the fluid inlet position in the absolute circumferential coordinate system.

When cavitation exists, α_c is determined by Equation 2.23 while $\pi < \alpha_c \leq 2\pi + \alpha_i$. Based on the predicted α_c, α_s is determined by Equation 2.24 while $(\alpha_c - 2\pi) \leq \alpha_s \leq \alpha_i$. Equations 2.23 and 2.24 are derived from Equations 1.37 and 1.38, respectively (Wang and Khonsari, 2008).

$$
\begin{aligned}
3\left(1-2\dot{\phi}\right)\varepsilon\left(1-\varepsilon^2\right)^{1/2}&\left[(2+\varepsilon\cos\alpha_c)(\sin\alpha_c-\sin\alpha_i)-\varepsilon\sin\alpha_i(\cos\alpha_c-\cos\alpha_i)\right.\\
&+(2\cos\alpha_c+\varepsilon)(\alpha_i-\alpha_c)\big]+6\dot{\varepsilon}\left[(2\cos\alpha_i-2\cos\alpha_c+\varepsilon\sin^2\alpha_i)(1-\varepsilon\cos\alpha_c)\right.\\
&-\varepsilon\sin\alpha_i\sin\alpha_c(4-\varepsilon\cos\alpha_i)-(2+\varepsilon^2)\sin\alpha_c(\alpha_c-\alpha_i)+3\varepsilon\sin^2\alpha_c\big]\\
&+\bar{p}_i\left(1-\varepsilon^2\right)^2(1-\varepsilon\cos\alpha_c)=0
\end{aligned}
\tag{2.23}
$$

$$
\begin{aligned}
3\left(1-2\dot{\phi}\right)\varepsilon\left(1-\varepsilon^2\right)^{1/2}&\left[2(\sin\alpha_s-\sin\alpha_i)(1+\varepsilon\cos\alpha_c)+(2\cos\alpha_c+\varepsilon)(\alpha_i-\alpha_s)\right.\\
&-\varepsilon(\sin\alpha_s\cos\alpha_s-\sin\alpha_i\cos\alpha_i)\big]+6\dot{\varepsilon}[4\varepsilon\sin\alpha_c(\sin\alpha_s-\sin\alpha_i)\\
&+(2-\varepsilon\cos\alpha_s-\varepsilon\cos\alpha_i)(\cos\alpha_i-\cos\alpha_s)(1-\varepsilon\cos\alpha_c)-(2+\varepsilon^2)\sin\alpha_c(\alpha_s-\alpha_i)\\
&-\varepsilon^2\sin\alpha_c(\sin\alpha_s\cos\alpha_s-\sin\alpha_i\cos\alpha_i)\big]+\bar{p}_i\left(1-\varepsilon^2\right)^2(1-\varepsilon\cos\alpha_c)=0
\end{aligned}
\tag{2.24}
$$

where the dimensionless fluid inlet pressure $\bar{p}_i = C^2/(\mu\omega R^2)p_i$, and "." represents $d/(\omega dt)$.

When cavitation does not exist, Equation 2.25 shall be used to determine α_c while $\pi < \alpha_c \leq 2\pi + \alpha_i$. Equation 2.25 is derived from Equation 1.39 given in Chapter 1 (Wang and Khonsari, 2008).

$$
\left(1-2\dot{\phi}\right)\varepsilon\left(1-\varepsilon^2\right)^{1/2}(2\cos\alpha_c+\varepsilon)+2\dot{\varepsilon}(2+\varepsilon^2)\sin\alpha_c=0
\tag{2.25}
$$

where "." represents $d/(\omega dt)$. Upon determining α_c, α_s equals $(\alpha_c - 2\pi)$.

Similarly, making use of Equations 2.20 and 2.21 for the fluid force components in long bearing, the two second-order equations of motion (Eqs. 2.3 and 2.4) are decomposed into four first-order equations. Letting $x_1 = \varepsilon$, $x_2 = \dot{\varepsilon}$, $x_3 = \phi$, $x_4 = \dot{\phi}$, $\Gamma_L = \mu L R^3/(2mC^{2.5}g^{0.5})$, and $\bar{\omega} = \omega\sqrt{C/g}$, the decomposed equations of motion are (Wang and Khonsari, 2008)

$$\dot{x}_1 = x_2$$

$$
\begin{aligned}
\dot{x}_2 = x_1 x_4^2 &+ \frac{1}{\bar{\omega}^2}\cos x_3\\
&-\frac{6\Gamma_L}{\bar{\omega}}\left[\frac{(1-2x_4)x_1(\cos\alpha_c-\cos\alpha_s)^2}{(1-x_1^2)(1-x_1\cos\alpha_c)}\right.\\
&\left.+2x_2\frac{(\alpha_c-\alpha_s+\sin\alpha_s\cos\alpha_s)(1-x_1\cos\alpha_c)+(x_1\cos^2\alpha_s-2\cos\alpha_s+\cos\alpha_c)\sin\alpha_c}{(1-x_1^2)^{3/2}(1-x_1\cos\alpha_c)}\right]
\end{aligned}
$$

$$\dot{x}_3 = x_4$$

$$\dot{x}_4 = -\frac{2x_2x_4}{x_1} - \frac{1}{\bar{\omega}^2 x_1}\sin x_3$$

$$+6\frac{\Gamma_L}{\bar{\omega}}\left\{\frac{(1-2x_4)}{\left(1-x_1^2\right)^{3/2}(1-x_1\cos\alpha_c)}\left[(1+2x_1\cos\alpha_c)(\alpha_c-\alpha_s)\right.\right.$$

$$-2(x_1+\cos\alpha_c)(\sin\alpha_c-\sin\alpha_s)+(\sin\alpha_c\cos\alpha_c-\sin\alpha_s\cos\alpha_s)]$$

$$+\frac{2x_2}{x_1\left(1-x_1^2\right)^2(1-x_1\cos\alpha_c)}\left[2x_1(\cos\alpha_c-x_1)-\left(2x_1\cos\alpha_s+\sin^2\alpha_s\right)(1-x_1\cos\alpha_c)\right.$$

$$\left.\left.-\sin^2\alpha_c+3x_1\sin\alpha_c(\alpha_c-\alpha_s)+\left(2x_1^2+2-x_1\cos\alpha_s\right)\sin\alpha_s\sin\alpha_c\right]\right\}$$

$$(2.26)$$

where g is the gravitational constant and Γ_L represents the characteristic constant for a specified long journal bearing assuming that the fluid viscosity μ is constant. Let parameter S represent the Sommerfeld number, then $\Gamma_L\bar{\omega}=(\pi/2)S$.

In Equation 2.26, α_c is defined by Equation 2.27 and α_s is given by Equation 2.28 when cavitation exists (Wang and Khonsari, 2008).

$$3(1-2x_4)x_1\left(1-x_1^2\right)^{1/2}[(2+x_1\cos\alpha_c)(\sin\alpha_c-\sin\alpha_i)-x_1\sin\alpha_i(\cos\alpha_c-\cos\alpha_i)$$

$$+(2\cos\alpha_c+x_1)(\alpha_i-\alpha_c)]+6x_2\left[(2\cos\alpha_i-2\cos\alpha_c+x_1\sin^2\alpha_i)(1-x_1\cos\alpha_c)\right.$$

$$-x_1\sin\alpha_i\sin\alpha_c(4-x_1\cos\alpha_i)-\left(2+x_1^2\right)\sin\alpha_c(\alpha_c-\alpha_i)+3x_1\sin^2\alpha_c]$$

$$+\bar{p}_i\left(1-x_1^2\right)^2(1-x_1\cos\alpha_c)=0$$

$$(2.27)$$

$$3(1-2x_4)x_1\left(1-x_1^2\right)^{1/2}[2(\sin\alpha_s-\sin\alpha_i)(1+x_1\cos\alpha_c)+(2\cos\alpha_c+x_1)(\alpha_i-\alpha_s)$$

$$-x_1(\sin\alpha_s\cos\alpha_s-\sin\alpha_i\cos\alpha_i)]+6x_2[4x_1\sin\alpha_c(\sin\alpha_s-\sin\alpha_i)$$

$$+(2-x_1\cos\alpha_s-x_1\cos\alpha_i)(\cos\alpha_i-\cos\alpha_s)(1-x_1\cos\alpha_c)-\left(2+x_1^2\right)\sin\alpha_c(\alpha_s-\alpha_i)$$

$$-x_1^2\sin\alpha_c(\sin\alpha_s\cos\alpha_s-\sin\alpha_i\cos\alpha_i)]+\bar{p}_i\left(1-x_1^2\right)^2(1-x_1\cos\alpha_c)=0$$

$$(2.28)$$

where α_i is given by Equation 2.29.

$$\cos\alpha_i=\frac{x_1+\cos(\Theta_i-x_3)}{1+x_1\cos(\Theta_i-x_3)}\qquad(2.29)$$

When cavitation does not exist, Equation 2.30 is used to determine α_c while $\pi < \alpha_c \leq 2\pi + \alpha_i$ (Wang and Khonsari, 2008).

$$(1 - 2x_4)x_1\left(1 - x_1^2\right)^{1/2}(2\cos\alpha_c + x_1) + 2x_2\left(2 + x_1^2\right)\sin\alpha_c = 0 \qquad (2.30)$$

Upon determining α_c, α_s equals $(\alpha_c - 2\pi)$.
The system of Equation 2.26 is of the form

$$\dot{\mathbf{x}} = \mathbf{f}(\mathbf{x}, \bar{\omega}) \qquad (2.31)$$

Similarly, the steady state equlibrium position $\mathbf{x}_s$ in terms of $(x_{1s} = \varepsilon_s,\ x_{2s} = \dot{\varepsilon}_s,\ x_{3s} = \phi_s,\ x_{4s} = \dot{\phi}_s)$ can be found analytically as $\mathbf{x}_s = (\varepsilon_s, 0, \phi_s, 0)$ by letting $\mathbf{f}(\mathbf{x}_s, \bar{\omega}) = 0$.

2.4 Summary

Equations 2.11 and 2.31 have demonstrated that irrespective of whether the rotor is supported by short bearings or long bearings, the rotor bearing system always has the same form of the decomposed equations of motion summarized by Equation 2.32.

$$\dot{\mathbf{x}} = \mathbf{f}(\mathbf{x}, \bar{\omega}) \qquad (2.32)$$

This universal form of the system equations of motion is suitable for numerical treatment using the Runge–Kutta–Fehlberg method. It also facilites the subsequent derivations of the analytical expressions for system instability threshold speed and whirl frequency. These topics are presented in Chapter 3.

References

Khonsari, M.M., Booser, E.R., 2008, *Applied Tribology: Bearing Design and Lubrication*, 2nd edition, John Wiley & Sons, New York.
Wang, J.K., Khonsari, M.M., 2006, "Application of Hopf Bifurcation Theory to the Rotor-Bearing System with Turbulent Effects," *Tribology International*, **39**, pp. 701–714.
Wang, J.K., Khonsari, M.M., 2008, "Effects of Oil Inlet Pressure and Inlet Position of axially grooved infinitely Long Journal Bearings. Part II: Nonlinear Instability Analysis," *Tribology International*, **41**, pp. 132–140.

3

Conventional Methods on System Instability Analysis

The importance of rotor-bearing system instability due to oil whirl in fluid-film journal bearings has been recognized since its discovery reported in 1925 by Newkirk and Taylor. While still not well understood, it remains to be an important design consideration in many types of modern rotating machinery.

Generally speaking, there are two different phenomena associated with oil-induced instability: oil whirl and oil whip. Oil whirl is a tendency of a journal to whirl moderately with a stable trajectory, while oil whip is a violent whipping motion triggered by a perturbation with its amplitude located outbound of the stability envelope (Wang and Khonsari, 2006a and 2006b). These two different phenomena appear to have been labelled indiscriminately in many of the existing literature (for example: Newkirk and Grobel, 1934; Pinkus, 1953; Newkirk and Lewis, 1956; Newkirk, 1956) that followed the classical work of Newkirk and Taylor (1925).

The classical paper of Newkirk and Taylor (1925) dealt with the oil whip of a rotor-bearing system. They pointed out that some rotor-bearing systems that were running quietly at speeds lower than the threshold speed began to whip violently just due a "special shock." Sometimes even a "slight" shock was sufficient to initiate violent whipping. Hori confirmed this phenomenon in 1959 and published extensive research on how, for example, an earthquake—which can be regarded as a shock—affects the occurrence of the oil whip of a rotor-bearing system (Hori, 1988; Hori and Kato, 1990). As for oil whirl, when the running speed of the rotor-bearing system is crossing the threshold speed, the journal will gradually lose

Thermohydrodynamic Instability in Fluid-Film Bearings, First Edition.
J. K. Wang and M. M. Khonsari.
© 2016 John Wiley & Sons, Ltd. Published 2016 by John Wiley & Sons, Ltd.

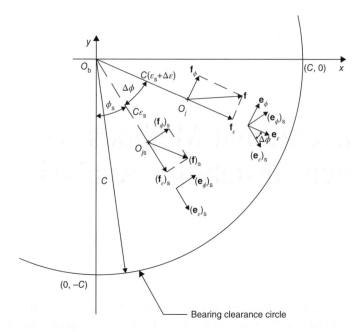

Figure 3.1 Decompositions of the fluid forces in journal bearing (Wang and Khonsari, 2006d)

its stability and the amplitude of the oil whirl (stable trajectory) will ramp up (Deepak and Noah, 1998; Wang and Khonsari, 2006a–c).

To evaluate the rotor-bearing system instability with fluid-film journal bearing involved, the equations of motion of the journal center must be solved. In the equations of motion (Eqs. 2.1 and 2.2), the hydrodynamic fluid force components (f_ε and f_ϕ) are highly nonlinear functions of both the position and velocity of the journal center. Figure 3.1 shows a sketch of the decompositions of the hydrodynamic fluid force **f** with a small perturbation ($\Delta\varepsilon$, $\Delta\phi$) from its steady-state position within a fluid-film journal bearing.

In Figure 3.1, $\mathbf{f}_\varepsilon$ and $\mathbf{f}_\phi$ are the radial and tangential components of the fluid forces **f** in journal bearing defined by Equations 2.13 and 2.14 in Section 2.2; $\mathbf{e}_\varepsilon$ and $\mathbf{e}_\phi$ are the unit vectors in the radial and tangential directions, respectively. Referring to Figure 3.1, the fluid force **f** is written in the following vector form (Wang and Khonsari, 2006d)

$$\mathbf{f} = f_\varepsilon \mathbf{e}_\varepsilon + f_\phi \mathbf{e}_\phi = \begin{bmatrix} \mathbf{e}_\varepsilon & \mathbf{e}_\phi \end{bmatrix} \begin{bmatrix} f_\varepsilon \\ f_\phi \end{bmatrix} \tag{3.1}$$

As shown in Sections 2.2 and 2.3, the hydrodynamic fluid force components are highly nonlinear. To evaluate the rotor-bearing stability, the highly nonlinear hydrodynamic fluid force components in the equations of motion (Eqs. 2.1 and

2.2) can be either linearized first and then solved analytically or solved directly and nonlinearly using a trial-and-error method.

3.1 Linearized Stiffness and Damping Method

3.1.1 Derivation of Linearized Bearing Stiffness and Damping Coefficients

Generally speaking, there are three commonly used approaches to derive expressions for the so-called stiffness and damping coefficients associated with the bearing fluid film. However, one of them leads to inconsistencies with the other two in the final expressions for the stiffness and damping coefficients. The inconsistencies become very obvious particularly when the derivations are applied to infinitely short bearings. The derivations of these coefficients and appropriate discussions are presented next (Wang and Khonsari, 2006d).

3.1.1.1 Two Approaches Based on Taylor Series Expansions

Two of the three commonly used definitions are based on the Taylor series expansions of the radial and tangential components ($\mathbf{f}_\varepsilon$ and $\mathbf{f}_\phi$) shown in Figure 3.1 individually and retaining only the zeroth-order and first-order terms (Wang and Khonsari, 2006d).

$$f_\varepsilon = \left(f_\varepsilon\right)_s + \begin{bmatrix} \dfrac{\partial f_\varepsilon}{\partial \varepsilon} & \dfrac{\partial f_\varepsilon}{\partial \phi} & \dfrac{\partial f_\varepsilon}{\partial \dot{\varepsilon}} & \dfrac{\partial f_\varepsilon}{\partial \dot{\phi}} \end{bmatrix} \begin{pmatrix} \Delta\varepsilon \\ \Delta\phi \\ \Delta\dot{\varepsilon} \\ \Delta\dot{\phi} \end{pmatrix}$$

$$f_\phi = \left(f_\phi\right)_s + \begin{bmatrix} \dfrac{\partial f_\phi}{\partial \varepsilon} & \dfrac{\partial f_\phi}{\partial \phi} & \dfrac{\partial f_\phi}{\partial \dot{\varepsilon}} & \dfrac{\partial f_\phi}{\partial \dot{\phi}} \end{bmatrix} \begin{pmatrix} \Delta\varepsilon \\ \Delta\phi \\ \Delta\dot{\varepsilon} \\ \Delta\dot{\phi} \end{pmatrix}$$

The above equations can be rearranged in the following form (Wang and Khonsari, 2006d):

$$\begin{pmatrix} f_\varepsilon - (f_\varepsilon)_s \\ f_\phi - (f_\phi)_s \end{pmatrix} = \begin{bmatrix} \dfrac{\partial f_\varepsilon}{C\partial\varepsilon} & \dfrac{\partial f_\varepsilon}{C\varepsilon\partial\phi} \\ \dfrac{\partial f_\phi}{C\partial\varepsilon} & \dfrac{\partial f_\phi}{C\varepsilon\partial\phi} \end{bmatrix} \begin{pmatrix} C\Delta\varepsilon \\ C\varepsilon\Delta\phi \end{pmatrix} + \begin{bmatrix} \dfrac{\partial f_\varepsilon}{C\partial\dot{\varepsilon}} & \dfrac{\partial f_\varepsilon}{C\varepsilon\partial\dot{\phi}} \\ \dfrac{\partial f_\phi}{C\partial\dot{\varepsilon}} & \dfrac{\partial f_\phi}{C\varepsilon\partial\dot{\phi}} \end{bmatrix} \begin{pmatrix} C\Delta\dot{\varepsilon} \\ C\varepsilon\Delta\dot{\phi} \end{pmatrix} \quad (3.2)$$

In Equation 3.2, all the first-order derivatives are evaluated at the steady-state equilibrium position (ε_s, ϕ_s).

Method 3.1   Direct Deduction from Taylor Expansion

Directly based on the Taylor series expansions (Eq. 3.2), one set of the stiffness coefficients (k_{ij}) and damping coefficients (b_{ij}) are defined as (Childs *et al.*, 1977):

$$\begin{pmatrix} f_\varepsilon - (f_\varepsilon)_s \\ f_\phi - (f_\phi)_s \end{pmatrix} = - \begin{bmatrix} k_{\varepsilon\varepsilon} & k_{\varepsilon\phi} \\ k_{\phi\varepsilon} & k_{\phi\phi} \end{bmatrix} \begin{pmatrix} C\Delta\varepsilon \\ C\varepsilon\Delta\phi \end{pmatrix} - \begin{bmatrix} b_{\varepsilon\varepsilon} & b_{\varepsilon\phi} \\ b_{\phi\varepsilon} & b_{\phi\phi} \end{bmatrix} \begin{pmatrix} C\Delta\dot\varepsilon \\ C\varepsilon\Delta\dot\phi \end{pmatrix} \qquad (3.3)$$

where

$$\begin{bmatrix} k_{\varepsilon\varepsilon} & k_{\varepsilon\phi} \\ k_{\phi\varepsilon} & k_{\phi\phi} \end{bmatrix} = \begin{bmatrix} -\dfrac{\partial f_\varepsilon}{C\partial\varepsilon} & -\dfrac{\partial f_\varepsilon}{C\varepsilon\partial\phi} \\[3mm] -\dfrac{\partial f_\phi}{C\partial\varepsilon} & -\dfrac{\partial f_\phi}{C\varepsilon\partial\phi} \end{bmatrix} \text{ and } \begin{bmatrix} b_{\varepsilon\varepsilon} & b_{\varepsilon\phi} \\ b_{\phi\varepsilon} & b_{\phi\phi} \end{bmatrix} = \begin{bmatrix} -\dfrac{\partial f_\varepsilon}{C\partial\dot\varepsilon} & -\dfrac{\partial f_\varepsilon}{C\varepsilon\partial\dot\phi} \\[3mm] -\dfrac{\partial f_\phi}{C\partial\dot\varepsilon} & -\dfrac{\partial f_\phi}{C\varepsilon\partial\dot\phi} \end{bmatrix} \qquad (3.4)$$

It is noteworthy to point out that, as defined in Figure 3.1, the hydrodynamic fluid force components f_ε and $(f_\varepsilon)_s$ in Equation 3.3 are in the different vector directions of $\mathbf{e}_\varepsilon$ and $(\mathbf{e}_\varepsilon)_s$, respectively. There is an angular rotation of $\Delta\phi$ from $(\mathbf{e}_\varepsilon)_s$ to $\mathbf{e}_\varepsilon$. Figure 3.1 also reveals that the vector directions of hydrodynamic fluid force components $\mathbf{f}_\phi$ and $(\mathbf{f}_\phi)_s$ have the same angular difference of $\Delta\phi$. The formulations of bearing stiffness and damping coefficients defined by Equations 3.3 and 3.4 are debatable due to the directional changes of the force components.

Method 3.2   Transformation Approach

As discussed earlier, there is an angular rotation of $\Delta\phi$ from $(\mathbf{e}_\varepsilon)_s$ to $\mathbf{e}_\varepsilon$. Based on Figure 3.1, Figure 3.2 adds another decomposition of the hydrodynamic fluid force $\mathbf{f}$ in the directions of steady-state unit vectors of $((\mathbf{e}_\varepsilon)_s$ and $(\mathbf{e}_\phi)_s)$.

From Figure 3.2, the following transformation is applied to reflect the angular rotation of $\Delta\phi$ between two different sets of decomposed force components (Holmes, 1960; Szeri, 1966).

$$\begin{pmatrix} f_\varepsilon{}' \\ f_\phi{}' \end{pmatrix} = \begin{pmatrix} \cos\Delta\phi & -\sin\Delta\phi \\ \sin\Delta\phi & \cos\Delta\phi \end{pmatrix} \begin{pmatrix} f_\varepsilon \\ f_\phi \end{pmatrix} \qquad (3.5)$$

where $\mathbf{f}_\varepsilon'$ is the component of the fluid force $\mathbf{f}$ in the same radial direction as the fluid force component $(\mathbf{f}_\varepsilon)_s$ at the steady-state equilibrium position, and $\mathbf{f}_\phi'$ is the component of the fluid force $\mathbf{f}$ in the same tangential direction as the fluid force component $(\mathbf{f}_\phi)_s$ at the steady-state equilibrium position.

For a small perturbation $(\Delta\phi \ll 1)$, $\cos\Delta\phi \approx 1$ and $\sin\Delta\phi \approx \Delta\phi$. Using these two approximations, Equation 3.5 can be rewritten as follows (Wang and Khonsari, 2006d):

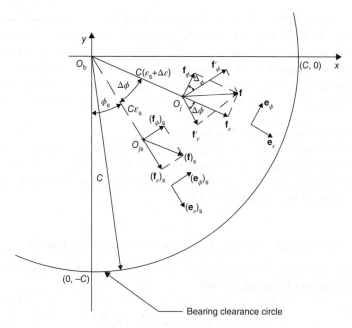

Figure 3.2 The decompositions of the fluid force **f** in different directions (Wang and Khonsari, 2006d)

$$\begin{pmatrix} f'_\varepsilon \\ f'_\phi \end{pmatrix} = \begin{pmatrix} f_\varepsilon \\ f_\phi \end{pmatrix} + \Delta\phi \begin{pmatrix} -f_\phi \\ f_\varepsilon \end{pmatrix} \tag{3.6}$$

Through combining and rearranging Equations 3.2 and 3.6, the following definitions of the stiffness and damping coefficients are developed (Holmes, 1960; Szeri, 1966).

$$\begin{pmatrix} f'_\varepsilon - (f_\varepsilon)_s \\ f'_\phi - (f_\phi)_s \end{pmatrix} = - \begin{bmatrix} k_{\varepsilon\varepsilon} & k_{\varepsilon\phi} \\ k_{\phi\varepsilon} & k_{\phi\phi} \end{bmatrix} \begin{pmatrix} C\Delta\varepsilon \\ C\varepsilon\Delta\phi \end{pmatrix} - \begin{bmatrix} b_{\varepsilon\varepsilon} & b_{\varepsilon\phi} \\ b_{\phi\varepsilon} & b_{\phi\phi} \end{bmatrix} \begin{pmatrix} C\Delta\dot{\varepsilon} \\ C\varepsilon\Delta\dot{\phi} \end{pmatrix} \tag{3.7}$$

where

$$\begin{bmatrix} k_{\varepsilon\varepsilon} & k_{\varepsilon\phi} \\ k_{\phi\varepsilon} & k_{\phi\phi} \end{bmatrix} = \begin{bmatrix} -\dfrac{\partial f_\varepsilon}{C\partial\varepsilon} & -\dfrac{\partial f_\varepsilon}{C\varepsilon\partial\phi} + \dfrac{f_\phi}{C\varepsilon} \\ -\dfrac{\partial f_\phi}{C\partial\varepsilon} & -\dfrac{\partial f_\phi}{C\varepsilon\partial\phi} - \dfrac{f_\varepsilon}{C\varepsilon} \end{bmatrix} \text{ and } \begin{bmatrix} b_{\varepsilon\varepsilon} & b_{\varepsilon\phi} \\ b_{\phi\varepsilon} & b_{\phi\phi} \end{bmatrix} = \begin{bmatrix} -\dfrac{\partial f_\varepsilon}{C\partial\dot{\varepsilon}} & -\dfrac{\partial f_\varepsilon}{C\varepsilon\partial\dot{\phi}} \\ -\dfrac{\partial f_\phi}{C\partial\dot{\varepsilon}} & -\dfrac{\partial f_\phi}{C\varepsilon\partial\dot{\phi}} \end{bmatrix}$$

$$\tag{3.8}$$

As defined in Figure 3.2, the hydrodynamic fluid force components $\mathbf{f}'_\varepsilon$ and $(\mathbf{f}_\varepsilon)_\text{s}$ in Equation 3.7 are in the same direction. So are the hydrodynamic fluid force components $\mathbf{f}'_\phi$ and $(\mathbf{f}_\phi)_\text{s}$. Therefore, the physical meaning of the bearing stiffness and damping coefficients defined by Equations 3.7 and 3.8 is more straightforward than those defined by Equations 3.3 and 3.4. They are more consistent with the conventional definitions of springs and dampeners (Wang and Khonsari, 2006d).

Descrepencies Between the Two Methods

Equations 3.4 and 3.8 show clear descrepencies of the definitions of $k_{\varepsilon\phi}$ and $k_{\phi\phi}$ derived from the two aforementioned methods. In the second definitions of $k_{\varepsilon\phi}$ and $k_{\phi\phi}$ defined by Equation 3.8, each item has one more coupled term than the first set of definitions defined by Equation 3.4.

3.1.1.2 Direct Approach Based on Vector Partial Differentiation

Taking the derivative of both sides of vector Equation 3.1 with respect to time t yields (Wang and Khonsari, 2006d):

$$\frac{d\mathbf{f}}{dt} = \frac{df_\varepsilon}{dt}\mathbf{e}_\varepsilon + \frac{df_\phi}{dt}\mathbf{e}_\phi + f_\varepsilon\frac{d\mathbf{e}_\varepsilon}{dt} + f_\phi\frac{d\mathbf{e}_\phi}{dt} \tag{3.9}$$

where

$$\begin{aligned}
\frac{df_\varepsilon}{dt} &= \frac{\partial f_\varepsilon}{\partial \varepsilon}\frac{d\varepsilon}{dt} + \frac{\partial f_\varepsilon}{\partial \phi}\frac{d\phi}{dt} + \frac{\partial f_\varepsilon}{\partial \dot{\varepsilon}}\frac{d\dot{\varepsilon}}{dt} + \frac{\partial f_\varepsilon}{\partial \dot{\phi}}\frac{d\dot{\phi}}{dt} \\
\frac{df_\phi}{dt} &= \frac{\partial f_\phi}{\partial \varepsilon}\frac{d\varepsilon}{dt} + \frac{\partial f_\phi}{\partial \phi}\frac{d\phi}{dt} + \frac{\partial f_\phi}{\partial \dot{\varepsilon}}\frac{d\dot{\varepsilon}}{dt} + \frac{\partial f_\phi}{\partial \dot{\phi}}\frac{d\dot{\phi}}{dt}
\end{aligned} \tag{3.10}$$

The derivatives of unit vectors $\mathbf{e}_\varepsilon$ and $\mathbf{e}_\phi$ are derived as (Greenwood, 1988):

$$\begin{aligned}
\frac{d\mathbf{e}_\varepsilon}{dt} &= \frac{d\phi}{dt}\mathbf{e}_\phi \\
\frac{d\mathbf{e}_\phi}{dt} &= -\frac{d\phi}{dt}\mathbf{e}_\varepsilon
\end{aligned} \tag{3.11}$$

Substituting Equations 3.10 and 3.11 into Equation 3.9, one has

$$\begin{aligned}
\frac{d\mathbf{f}}{dt} = &\left(\frac{\partial f_\varepsilon}{\partial \varepsilon}\frac{d\varepsilon}{dt} + \frac{\partial f_\varepsilon}{\partial \phi}\frac{d\phi}{dt} + \frac{\partial f_\varepsilon}{\partial \dot{\varepsilon}}\frac{d\dot{\varepsilon}}{dt} + \frac{\partial f_\varepsilon}{\partial \dot{\phi}}\frac{d\dot{\phi}}{dt} - \frac{d\phi}{dt}f_\phi \right)\mathbf{e}_\varepsilon \\
&+ \left(\frac{\partial f_\phi}{\partial \varepsilon}\frac{d\varepsilon}{dt} + \frac{\partial f_\phi}{\partial \phi}\frac{d\phi}{dt} + \frac{\partial f_\phi}{\partial \dot{\varepsilon}}\frac{d\dot{\varepsilon}}{dt} + \frac{\partial f_\phi}{\partial \dot{\phi}}\frac{d\dot{\phi}}{dt} + \frac{d\phi}{dt}f_\varepsilon \right)\mathbf{e}_\phi
\end{aligned}$$

Simplifying this expression further yields

$$
\frac{d\mathbf{f}}{dt} = \begin{bmatrix} \mathbf{e}_\varepsilon & \mathbf{e}_\phi \end{bmatrix}
\begin{bmatrix}
\dfrac{\partial f_\varepsilon}{\partial \varepsilon}\dfrac{d\varepsilon}{dt} + \left(\dfrac{\partial f_\varepsilon}{\partial \phi} - f_\phi \right)\dfrac{d\phi}{dt} + \dfrac{\partial f_\varepsilon}{\partial \dot\varepsilon}\dfrac{d\dot\varepsilon}{dt} + \dfrac{\partial f_\varepsilon}{\partial \dot\phi}\dfrac{d\dot\phi}{dt} \\[2ex]
\dfrac{\partial f_\phi}{\partial \varepsilon}\dfrac{d\varepsilon}{dt} + \left(\dfrac{\partial f_\phi}{\partial \phi} + f_\varepsilon \right)\dfrac{d\phi}{dt} + \dfrac{\partial f_\phi}{\partial \dot\varepsilon}\dfrac{d\dot\varepsilon}{dt} + \dfrac{\partial f_\phi}{\partial \dot\phi}\dfrac{d\dot\phi}{dt}
\end{bmatrix}
\tag{3.12}
$$

Equation 3.12 can be approximated by multiplying both sides by Δt as shown in Equation 3.13. The expression in Equation 3.13 describes the variation of fluid force with time when a small perturbation for a very short time interval Δt is applied to the journal that is at steady-state equilibrium position (ε_s, ϕ_s) (Wang and Khonsari, 2006d).

$$
\mathbf{f} - \mathbf{f}_s = \begin{bmatrix} \mathbf{e}_\varepsilon & \mathbf{e}_\phi \end{bmatrix}
\begin{bmatrix}
\dfrac{\partial f_\varepsilon}{\partial \varepsilon}\Delta\varepsilon + \left(\dfrac{\partial f_\varepsilon}{\partial \phi} - f_\phi \right)\Delta\phi + \dfrac{\partial f_\varepsilon}{\partial \dot\varepsilon}\Delta\dot\varepsilon + \dfrac{\partial f_\varepsilon}{\partial \dot\phi}\Delta\dot\phi \\[2ex]
\dfrac{\partial f_\phi}{\partial \varepsilon}\Delta\varepsilon + \left(\dfrac{\partial f_\phi}{\partial \phi} + f_\varepsilon \right)\Delta\phi + \dfrac{\partial f_\phi}{\partial \dot\varepsilon}\Delta\dot\varepsilon + \dfrac{\partial f_\phi}{\partial \dot\phi}\Delta\dot\phi
\end{bmatrix}
\tag{3.13}
$$

Here, the fluid force at the steady-state equilibrium position (ε_s, ϕ_s) is given by

$$
\mathbf{f}_s = \begin{bmatrix} \mathbf{e}_\varepsilon & \mathbf{e}_\phi \end{bmatrix}
\begin{bmatrix}
(f_\varepsilon)_s \\[1ex]
(f_\phi)_s
\end{bmatrix}
\tag{3.14}
$$

Substituting Equations 3.1 and 3.14 into Equation 3.13 and rearranging it yields

$$
\begin{bmatrix} \mathbf{e}_\varepsilon & \mathbf{e}_\phi \end{bmatrix}
\begin{bmatrix}
f_\varepsilon - (f_\varepsilon)_s \\[1ex]
f_\phi - (f_\phi)_s
\end{bmatrix}
= \begin{bmatrix} \mathbf{e}_\varepsilon & \mathbf{e}_\phi \end{bmatrix}
\begin{bmatrix}
\dfrac{\partial f_\varepsilon}{\partial \varepsilon}\Delta\varepsilon + \left(\dfrac{\partial f_\varepsilon}{\partial \phi} - f_\phi \right)\Delta\phi + \dfrac{\partial f_\varepsilon}{\partial \dot\varepsilon}\Delta\dot\varepsilon + \dfrac{\partial f_\varepsilon}{\partial \dot\phi}\Delta\dot\phi \\[2ex]
\dfrac{\partial f_\phi}{\partial \varepsilon}\Delta\varepsilon + \left(\dfrac{\partial f_\phi}{\partial \phi} + f_\varepsilon \right)\Delta\phi + \dfrac{\partial f_\phi}{\partial \dot\varepsilon}\Delta\dot\varepsilon + \dfrac{\partial f_\phi}{\partial \dot\phi}\Delta\dot\phi
\end{bmatrix}
$$

That is,

$$
\begin{bmatrix}
f_\varepsilon - (f_\varepsilon)_s \\[1ex]
f_\phi - (f_\phi)_s
\end{bmatrix}
= \begin{bmatrix}
\dfrac{\partial f_\varepsilon}{\partial \varepsilon}\Delta\varepsilon + \left(\dfrac{\partial f_\varepsilon}{\partial \phi} - f_\phi \right)\Delta\phi + \dfrac{\partial f_\varepsilon}{\partial \dot\varepsilon}\Delta\dot\varepsilon + \dfrac{\partial f_\varepsilon}{\partial \dot\phi}\Delta\dot\phi \\[2ex]
\dfrac{\partial f_\phi}{\partial \varepsilon}\Delta\varepsilon + \left(\dfrac{\partial f_\phi}{\partial \phi} + f_\varepsilon \right)\Delta\phi + \dfrac{\partial f_\phi}{\partial \dot\varepsilon}\Delta\dot\varepsilon + \dfrac{\partial f_\phi}{\partial \dot\phi}\Delta\dot\phi
\end{bmatrix}
\tag{3.15}
$$

Equation 3.15 can be rewritten as follows (Wang and Khonsari, 2006d):

$$
\begin{bmatrix} f_\varepsilon - (f_\varepsilon)_s \\ f_\phi - (f_\phi)_s \end{bmatrix} = \begin{bmatrix} \dfrac{\partial f_\varepsilon}{C\partial\varepsilon} & \dfrac{\partial f_\varepsilon}{C\varepsilon\partial\phi} - \dfrac{f_\phi}{C\varepsilon} \\ \dfrac{\partial f_\phi}{C\partial\varepsilon} & \dfrac{\partial f_\phi}{C\varepsilon\partial\phi} + \dfrac{f_\varepsilon}{C\varepsilon} \end{bmatrix} \begin{bmatrix} C\Delta\varepsilon \\ C\varepsilon\Delta\phi \end{bmatrix} + \begin{bmatrix} \dfrac{\partial f_\varepsilon}{C\partial\dot{\varepsilon}} & \dfrac{\partial f_\varepsilon}{C\varepsilon\partial\dot{\phi}} \\ \dfrac{\partial f_\phi}{C\partial\dot{\varepsilon}} & \dfrac{\partial f_\phi}{C\varepsilon\partial\dot{\phi}} \end{bmatrix} \begin{bmatrix} C\Delta\dot{\varepsilon} \\ C\varepsilon\Delta\dot{\phi} \end{bmatrix} \tag{3.16}
$$

Based on Equation 3.16, the stiffness and damping coefficients can be defined as follows (Wang and Khonsari, 2006d):

$$
\begin{bmatrix} f_\varepsilon - (f_\varepsilon)_s \\ f_\phi - (f_\phi)_s \end{bmatrix} = - \begin{bmatrix} k_{\varepsilon\varepsilon} & k_{\varepsilon\phi} \\ k_{\phi\varepsilon} & k_{\phi\phi} \end{bmatrix} \begin{bmatrix} C\Delta\varepsilon \\ C\varepsilon\Delta\phi \end{bmatrix} - \begin{bmatrix} b_{\varepsilon\varepsilon} & b_{\varepsilon\phi} \\ b_{\phi\varepsilon} & b_{\phi\phi} \end{bmatrix} \begin{bmatrix} C\Delta\dot{\varepsilon} \\ C\varepsilon\Delta\dot{\phi} \end{bmatrix} \tag{3.17}
$$

where

$$
\begin{bmatrix} k_{\varepsilon\varepsilon} & k_{\varepsilon\phi} \\ k_{\phi\varepsilon} & k_{\phi\phi} \end{bmatrix} = \begin{bmatrix} -\dfrac{\partial f_\varepsilon}{C\partial\varepsilon} & -\dfrac{\partial f_\varepsilon}{C\varepsilon\partial\phi} + \dfrac{f_\phi}{C\varepsilon} \\ -\dfrac{\partial f_\phi}{C\partial\varepsilon} & -\dfrac{\partial f_\phi}{C\varepsilon\partial\phi} - \dfrac{f_\varepsilon}{C\varepsilon} \end{bmatrix} \text{ and } \begin{bmatrix} b_{\varepsilon\varepsilon} & b_{\varepsilon\phi} \\ b_{\phi\varepsilon} & b_{\phi\phi} \end{bmatrix} = \begin{bmatrix} -\dfrac{\partial f_\varepsilon}{C\partial\dot{\varepsilon}} & -\dfrac{\partial f_\varepsilon}{C\varepsilon\partial\dot{\phi}} \\ -\dfrac{\partial f_\phi}{C\partial\dot{\varepsilon}} & -\dfrac{\partial f_\phi}{C\varepsilon\partial\dot{\phi}} \end{bmatrix}
$$

$$
\tag{3.18}
$$

The stiffness and damping coefficients defined by Equations 3.18 are consistent with Equation 3.8 in Method 3.2. However, the physical meaning in this derivation is clearer and simpler. In Equation 3.17, the change of the fluid force component $\mathbf{f}_\varepsilon$ is exactly the relative force change in the direction of $\mathbf{e}_\varepsilon$ due to the small perturbation ($\Delta\varepsilon$, $\Delta\phi$, $\Delta\dot{\varepsilon}$, $\Delta\dot{\phi}$), and the change of the fluid force component $\mathbf{f}_\phi$ is exactly the relative force change in the direction of $\mathbf{e}_\phi$ due to the small perturbation ($\Delta\varepsilon$, $\Delta\phi$, $\Delta\dot{\varepsilon}$, $\Delta\dot{\phi}$) (Wang and Khonsari, 2006d).

3.1.1.3 Verification and Discussion

As an example, the definitions of stiffness and damping coefficients described by Equations 3.17 and 3.18 are applied to short bearings in this section. Referring to Section 1.2, assuming that the short-bearing theory with half-Sommerfeld boundary conditions applies, the fluid forces in the journal bearing are given by

$$
f_\varepsilon = -\frac{RL^3\mu}{2C^2}\left[\frac{2\varepsilon^2(\omega-2\dot{\phi})}{(1-\varepsilon^2)^2} + \frac{\pi(1+2\varepsilon^2)\dot{\varepsilon}}{(1-\varepsilon^2)^{5/2}}\right] \tag{3.19}
$$

$$
f_\phi = \frac{RL^3\mu}{2C^2}\left[\frac{\pi(\omega-2\dot{\phi})\varepsilon}{2(1-\varepsilon^2)^{3/2}} + \frac{4\varepsilon\dot{\varepsilon}}{(1-\varepsilon^2)^2}\right] \tag{3.20}
$$

Substituting Equations 3.19 and 3.20 into Equation 3.18 and simplifying the resulting expressions, the linearized stiffness coefficients k_{ij} $(i,j = \varepsilon, \phi)$ for short bearings are derived as follows (Wang and Khonsari, 2006d):

$$k_{\varepsilon\varepsilon} = \frac{2\omega\mu R L^3 \varepsilon (1 + \varepsilon^2)}{C^3 (1 - \varepsilon^2)^3}$$

$$k_{\varepsilon\phi} = \frac{\pi\omega\mu R L^3}{4C^3 (1 - \varepsilon^2)^{3/2}}$$

$$k_{\phi\varepsilon} = -\frac{\pi\omega\mu R L^3 (1 + 2\varepsilon^2)}{4C^3 (1 - \varepsilon^2)^{5/2}}$$

$$k_{\phi\phi} = \frac{\varepsilon\omega\mu R L^3}{C^3 (1 - \varepsilon^2)^2}$$

(3.21)

The damping coefficients b_{ij} $(i,j = \varepsilon, \phi)$ for short bearings are given as follows (Wang and Khonsari, 2006d):

$$b_{\varepsilon\varepsilon} = \frac{\pi\mu R L^3 (1 + 2\varepsilon^2)}{2C^3 (1 - \varepsilon^2)^{5/2}}$$

$$b_{\varepsilon\phi} = -\frac{2\varepsilon\mu R L^3}{C^3 (1 - \varepsilon^2)^2}$$

$$b_{\phi\varepsilon} = -\frac{2\varepsilon\mu R L^3}{C^3 (1 - \varepsilon^2)^2}$$

$$b_{\phi\phi} = \frac{\pi\mu R L^3}{2C^3 (1 - \varepsilon^2)^{3/2}}$$

(3.22)

The above linearized stiffness and damping coefficients can be normalized by $\bar{k}_{ij} = \frac{(C/R)^3}{\mu\omega L} k_{ij}$, $(i,j = \varepsilon, \phi)$ and $\bar{b}_{ij} = \frac{(C/R)^3}{\mu L} b_{ij}$, $(i,j = \varepsilon, \phi)$. The normalized expressions for the stiffness and damping coefficients agree with those in references (Holmes, 1960; Szeri, 1966). However, if the definitions described by Equations 3.3 and 3.4 are applied to the same infinitely short-bearing theory, $k_{\varepsilon\phi} \equiv 0$ and $k_{\phi\phi} \equiv 0$ since both of the fluid force components f_ε and f_ϕ are not explicit functions of ϕ. This example shows that the stiffness and damping coefficients defined by Equations 3.8 and 3.18 are more suitable to the application of fluid film journal bearings than those defined by Equation 3.4 (Wang and Khonsari, 2006d).

In summary, the direct approach based on hydrodynamic force vector partial differentiation presents a simple procedure to derive the stiffness and damping coefficients in polar coordinates. This method also offers a clear physical meaning of the hydrodynamic stiffness and damping coefficients and yields consistent results especially when applied to infinitely short bearings (Wang and Khonsari, 2006d).

3.1.2   Instability Threshold Speed Based on the Linearized Stiffness and Damping Coefficients

Referring to the equations of motion (Eq. 2.32) derived in Section 2.3, we now present an expression for determining the instability threshold speed based on the linearized stiffness and damping coefficients derived in the previous section.

The Jacobian matrix of the equations of motion (2.32) at the stationary point $\mathbf{x}_s$ $(\varepsilon_s, 0, \phi_s, 0)$ where $\mathbf{f}(\mathbf{x}_s, \bar{\omega}) = 0$ is given as (Wang and Khonsari, 2006e):

$$\mathbf{f}_\mathbf{x}(\mathbf{x}_s, \omega) = \frac{\partial \mathbf{f}}{\partial \mathbf{x}}(\mathbf{x}_s, \omega)$$

$$= \begin{bmatrix} 0 & 1 & 0 & 0 \\[2mm] \dfrac{1}{m\omega^2}\dfrac{\partial f_\varepsilon}{C\partial x_1} & \dfrac{1}{m\omega^2}\dfrac{\partial f_\varepsilon}{C\partial x_2} & -\dfrac{x_1}{m\omega^2}\left(-\dfrac{\partial f_\varepsilon}{Cx_1\partial x_3} + \dfrac{f_\phi}{Cx_1}\right) & \dfrac{x_1}{m\omega^2}\dfrac{\partial f_\varepsilon}{Cx_1\partial x_4} \\[2mm] 0 & 0 & 0 & 1 \\[2mm] \dfrac{1}{x_1 m\omega^2}\dfrac{\partial f_\phi}{C\partial x_1} & \dfrac{1}{x_1 m\omega^2}\dfrac{\partial f_\phi}{C\partial x_2} & \dfrac{1}{m\omega^2}\left(\dfrac{\partial f_\phi}{Cx_1\partial x_3} + \dfrac{f_\varepsilon}{Cx_1}\right) & \dfrac{1}{m\omega^2}\dfrac{\partial f_\phi}{Cx_1\partial x_4} \end{bmatrix}$$

$$(3.23)$$

In terms of the linearized dimensional stiffness and damping coefficients given by Equation 3.18 and making use of the definitions of $\bar{t} = \omega t$, $x_1 = \varepsilon$, $x_2 = \dot{\varepsilon}$, $x_3 = \phi$ and $x_4 = \dot{\phi}$, Equation 3.23 can be rewritten as follows (Wang and Khonsari, 2006e):

$$\mathbf{f}_\mathbf{x}(\mathbf{x}_s, \omega) = \begin{bmatrix} 0 & 1 & 0 & 0 \\[2mm] -\dfrac{1}{m\omega^2}k_{\varepsilon\varepsilon} & -\dfrac{1}{m\omega}b_{\varepsilon\varepsilon} & -\dfrac{\varepsilon}{m\omega^2}k_{\varepsilon\phi} & -\dfrac{\varepsilon}{m\omega}b_{\varepsilon\phi} \\[2mm] 0 & 0 & 0 & 1 \\[2mm] -\dfrac{1}{\varepsilon m\omega^2}k_{\phi\varepsilon} & -\dfrac{1}{\varepsilon m\omega}b_{\phi\varepsilon} & -\dfrac{1}{m\omega^2}k_{\phi\phi} & -\dfrac{1}{m\omega}b_{\phi\phi} \end{bmatrix}$$

$$(3.24)$$

The Jacobian matrix can be normalized by letting $\bar{k}_{ij} = (C/R)^3/(\mu\omega L)k_{ij}$ while $(i,j = \varepsilon, \phi)$, $\bar{b}_{ij} = (C/R)^3/(\mu L)b_{ij}$ while $(i,j = \varepsilon, \phi)$, $S = \mu\omega LD/(2\pi mg)(R/C)^2$, and $\bar{\omega} = \omega\sqrt{C/g}$. It leads to the following form (Wang and Khonsari, 2006e):

$$\mathbf{f}_\mathbf{x}(\mathbf{x}_s, \bar{\omega}) = \begin{bmatrix} 0 & 1 & 0 & 0 \\[2mm] -\dfrac{\pi S}{\bar{\omega}^2}\bar{k}_{\varepsilon\varepsilon} & -\dfrac{\pi S}{\bar{\omega}^2}\bar{b}_{\varepsilon\varepsilon} & -\dfrac{\varepsilon\pi S}{\bar{\omega}^2}\bar{k}_{\varepsilon\phi} & -\dfrac{\varepsilon\pi S}{\bar{\omega}^2}\bar{b}_{\varepsilon\phi} \\[2mm] 0 & 0 & 0 & 1 \\[2mm] -\dfrac{\pi S}{\varepsilon\bar{\omega}^2}\bar{k}_{\phi\varepsilon} & -\dfrac{\pi S}{\varepsilon\bar{\omega}^2}\bar{b}_{\phi\varepsilon} & -\dfrac{\pi S}{\bar{\omega}^2}\bar{k}_{\phi\phi} & -\dfrac{\pi S}{\bar{\omega}^2}\bar{b}_{\phi\phi} \end{bmatrix}$$

$$(3.25)$$

For a very small interval $\Delta \mathbf{x}$, based on the definition of partial differential and at the steady state $\mathbf{f}(\mathbf{x}_s, \bar{\omega}) = 0$, multiplying the both sides of Equation 3.25 by $\Delta \mathbf{x}$ leads to

$$\mathbf{f}_{\mathbf{x}}(\mathbf{x}_s, \bar{\omega}) \Delta \mathbf{x} = \Delta \mathbf{f}(\mathbf{x}_s, \bar{\omega}) = \mathbf{f}(\mathbf{x}, \bar{\omega}) - \mathbf{f}(\mathbf{x}_s, \bar{\omega}) = \mathbf{f}(\mathbf{x}, \bar{\omega})$$

Substituting the equation of motion $\dot{\mathbf{x}} = \mathbf{f}(\mathbf{x}, \bar{\omega})$ into the above equation, one arrives at:

$$\begin{pmatrix} \dot{\varepsilon} \\ \ddot{\varepsilon} \\ \dot{\phi} \\ \ddot{\phi} \end{pmatrix} = \begin{bmatrix} 0 & 1 & 0 & 0 \\ -\dfrac{\pi S}{\bar{\omega}^2}\bar{k}_{\varepsilon\varepsilon} & -\dfrac{\pi S}{\bar{\omega}^2}\bar{b}_{\varepsilon\varepsilon} & -\dfrac{\varepsilon\pi S}{\bar{\omega}^2}\bar{k}_{\varepsilon\phi} & -\dfrac{\varepsilon\pi S}{\bar{\omega}^2}\bar{b}_{\varepsilon\phi} \\ 0 & 0 & 0 & 1 \\ -\dfrac{\pi S}{\varepsilon\bar{\omega}^2}\bar{k}_{\phi\varepsilon} & -\dfrac{\pi S}{\varepsilon\bar{\omega}^2}\bar{b}_{\phi\varepsilon} & -\dfrac{\pi S}{\bar{\omega}^2}\bar{k}_{\phi\phi} & -\dfrac{\pi S}{\bar{\omega}^2}\bar{b}_{\phi\phi} \end{bmatrix} \begin{pmatrix} \Delta\varepsilon \\ \Delta\dot{\varepsilon} \\ \Delta\phi \\ \Delta\dot{\phi} \end{pmatrix}$$

Extracting the second and fourth equations from the above equation matrix results in (Wang and Khonsari, 2006e)

$$\begin{pmatrix} \ddot{\varepsilon} \\ \ddot{\phi} \end{pmatrix} = \begin{bmatrix} -\dfrac{\pi S}{\bar{\omega}^2}\bar{k}_{\varepsilon\varepsilon} & -\dfrac{\pi S}{\bar{\omega}^2}\bar{b}_{\varepsilon\varepsilon} & -\dfrac{\varepsilon\pi S}{\bar{\omega}^2}\bar{k}_{\varepsilon\phi} & -\dfrac{\varepsilon\pi S}{\bar{\omega}^2}\bar{b}_{\varepsilon\phi} \\ -\dfrac{\pi S}{\varepsilon\bar{\omega}^2}\bar{k}_{\phi\varepsilon} & -\dfrac{\pi S}{\varepsilon\bar{\omega}^2}\bar{b}_{\phi\varepsilon} & -\dfrac{\pi S}{\bar{\omega}^2}\bar{k}_{\phi\phi} & -\dfrac{\pi S}{\bar{\omega}^2}\bar{b}_{\phi\phi} \end{bmatrix} \begin{pmatrix} \Delta\varepsilon \\ \Delta\dot{\varepsilon} \\ \Delta\phi \\ \Delta\dot{\phi} \end{pmatrix} \qquad (3.26)$$

If the system is running at the instability threshold speed $\bar{\omega}_{st}$, then the orbit of the journal will take on the following form (Majumdar *et al.*, 1988):

$$\varepsilon = \varepsilon_s + A_\varepsilon e^{i\omega_s t} = \varepsilon_s + A_\varepsilon e^{i\Omega \bar{t}}$$
$$\phi = \phi_s + A_\phi e^{i\omega_s t} = \phi_s + A_\phi e^{i\Omega \bar{t}} \qquad (3.27)$$

where $\Omega = \bar{\omega}_s / \bar{\omega}_{st}$ is the whirling frequency ratio at the threshold speed $\bar{\omega}_{st}$. Substituting Equation 3.27 into Equation 3.26, Equation 3.28 is obtained.

$$A_\varepsilon \left(\Omega^2 - i\Omega \dfrac{\pi S}{\bar{\omega}_{st}^2}\bar{b}_{\varepsilon\varepsilon} - \dfrac{\pi S}{\bar{\omega}_{st}^2}\bar{k}_{\varepsilon\varepsilon} \right) + A_\phi \left(-i\Omega \dfrac{\pi\varepsilon_s S}{\bar{\omega}_{st}^2}\bar{b}_{\varepsilon\phi} - \dfrac{\pi\varepsilon_s S}{\bar{\omega}_{st}^2}\bar{k}_{\varepsilon\phi} \right) = 0$$
$$A_\varepsilon \left(-i\Omega \dfrac{\pi S}{\varepsilon_s\bar{\omega}_{st}^2}\bar{b}_{\phi\varepsilon} - \dfrac{\pi S}{\varepsilon_s\bar{\omega}_{st}^2}\bar{k}_{\phi\varepsilon} \right) + A_\phi \left(\Omega^2 - i\Omega \dfrac{\pi S}{\bar{\omega}_{st}^2}\bar{b}_{\phi\phi} - \dfrac{\pi S}{\bar{\omega}_{st}^2}\bar{k}_{\phi\phi} \right) = 0 \qquad (3.28)$$

To ensure a nontrivial solution to Equation 3.28, the following condition must be satisfied (Wang and Khonsari, 2006e):

$$
\begin{vmatrix}
\Omega^2 - i\Omega\dfrac{\pi S}{\bar{\omega}_{st}^2}\bar{b}_{\varepsilon\varepsilon} - \dfrac{\pi S}{\bar{\omega}_{st}^2}\bar{k}_{\varepsilon\varepsilon} & -i\Omega\dfrac{\pi\varepsilon_s S}{\bar{\omega}_{st}^2}\bar{b}_{\varepsilon\phi} - \dfrac{\pi\varepsilon_s S}{\bar{\omega}_{st}^2}\bar{k}_{\varepsilon\phi} \\[2mm]
-i\Omega\dfrac{\pi S}{\varepsilon_s\bar{\omega}_{st}^2}\bar{b}_{\phi\varepsilon} - \dfrac{\pi S}{\varepsilon_s\bar{\omega}_{st}^2}\bar{k}_{\phi\varepsilon} & \Omega^2 - i\Omega\dfrac{\pi S}{\bar{\omega}_{st}^2}\bar{b}_{\phi\phi} - \dfrac{\pi S}{\bar{\omega}_{st}^2}\bar{k}_{\phi\phi}
\end{vmatrix} = 0 \qquad (3.29)
$$

That is

$$
\begin{aligned}
&\left(\bar{\omega}_{st}^2\Omega^2 - i\Omega\pi S\bar{b}_{\varepsilon\varepsilon} - \pi S\bar{k}_{\varepsilon\varepsilon}\right)\left(\bar{\omega}_{st}^2\Omega^2 - i\Omega\pi S\bar{b}_{\phi\phi} - \pi S\bar{k}_{\phi\phi}\right) \\
&- \left(i\Omega\pi S\bar{b}_{\phi\varepsilon} + \pi S\bar{k}_{\phi\varepsilon}\right)\left(i\Omega\pi S\bar{b}_{\varepsilon\phi} + \pi S\bar{k}_{\varepsilon\phi}\right) = 0
\end{aligned} \qquad (3.30)
$$

Let $\bar{K}_{ij} = \pi S\bar{k}_{ij}$, $\bar{B}_{ij} = \pi S\bar{b}_{ij}$ $(i,j = \varepsilon,\phi)$, Equation 3.30 can be rewritten as follows:

$$
\begin{aligned}
&\left(\bar{\omega}_{st}^2\Omega^2 - \bar{K}_{\varepsilon\varepsilon}\right)\left(\bar{\omega}_{st}^2\Omega^2 - \bar{K}_{\phi\phi}\right) - \Omega^2\bar{B}_{\varepsilon\varepsilon}\bar{B}_{\phi\phi} - \bar{K}_{\phi\varepsilon}\bar{K}_{\varepsilon\phi} + \Omega^2\bar{B}_{\varepsilon\phi}\bar{B}_{\phi\varepsilon} \\
&- i\Omega\left[\bar{B}_{\phi\phi}\left(\bar{\omega}_{st}^2\Omega^2 - \bar{K}_{\varepsilon\varepsilon}\right) + \bar{B}_{\varepsilon\varepsilon}\left(\bar{\omega}_{st}^2\Omega^2 - \bar{K}_{\phi\phi}\right) + \bar{B}_{\phi\varepsilon}\bar{K}_{\varepsilon\phi} + \bar{B}_{\varepsilon\phi}\bar{K}_{\phi\varepsilon}\right] = 0
\end{aligned} \qquad (3.31)
$$

Letting the sum of the real terms and the sum of the imaginary terms of the left hand side of Equation 3.31 equal to 0, respectively, yields:

$$
\left(\bar{\omega}_{st}^2\Omega^2 - \bar{K}_{\varepsilon\varepsilon}\right)\left(\bar{\omega}_{st}^2\Omega^2 - \bar{K}_{\phi\phi}\right) - \Omega^2\bar{B}_{\varepsilon\varepsilon}\bar{B}_{\phi\phi} - \bar{K}_{\phi\varepsilon}\bar{K}_{\varepsilon\phi} + \Omega^2\bar{B}_{\varepsilon\phi}\bar{B}_{\phi\varepsilon} = 0
$$

$$
\bar{B}_{\phi\phi}\left(\bar{\omega}_{st}^2\Omega^2 - \bar{K}_{\varepsilon\varepsilon}\right) + \bar{B}_{\varepsilon\varepsilon}\left(\bar{\omega}_{st}^2\Omega^2 - \bar{K}_{\phi\phi}\right) + \bar{B}_{\phi\varepsilon}\bar{K}_{\varepsilon\phi} + \bar{B}_{\varepsilon\phi}\bar{K}_{\phi\varepsilon} = 0
$$

Through combining and rearranging these two equations, we obtain the following relationships (Wang and Khonsari, 2006e).

$$
\bar{\omega}_s^2 = \bar{\omega}_{st}^2\Omega^2 = \frac{\bar{K}_{\varepsilon\varepsilon}\bar{B}_{\phi\phi} + \bar{K}_{\phi\phi}\bar{B}_{\varepsilon\varepsilon} - \bar{K}_{\varepsilon\phi}\bar{B}_{\phi\varepsilon} - \bar{K}_{\phi\varepsilon}\bar{B}_{\varepsilon\phi}}{\bar{B}_{\varepsilon\varepsilon} + \bar{B}_{\phi\phi}} \qquad (3.32)
$$

$$
\Omega^2 = \frac{\bar{\omega}_s^2}{\bar{\omega}_{st}^2} = \frac{\left(\bar{\omega}_s^2 - \bar{K}_{\varepsilon\varepsilon}\right)\left(\bar{\omega}_s^2 - \bar{K}_{\phi\phi}\right) - \bar{K}_{\varepsilon\phi}\bar{K}_{\phi\varepsilon}}{\bar{B}_{\varepsilon\varepsilon}\bar{B}_{\phi\phi} - \bar{B}_{\varepsilon\phi}\bar{B}_{\phi\varepsilon}} \qquad (3.33)
$$

With the rearrangement of Equation 3.33, the dimensionless threshold speed is defined as (Wang and Khonsari, 2006e)

$$
\bar{\omega}_{st} = \bar{\omega}_s\sqrt{\frac{\bar{B}_{\varepsilon\varepsilon}\bar{B}_{\phi\phi} - \bar{B}_{\varepsilon\phi}\bar{B}_{\phi\varepsilon}}{\left(\bar{\omega}_s^2 - \bar{K}_{\varepsilon\varepsilon}\right)\left(\bar{\omega}_s^2 - \bar{K}_{\phi\phi}\right) - \bar{K}_{\varepsilon\phi}\bar{K}_{\phi\varepsilon}}} \qquad (3.34)
$$

where the dimensionless whirl speed is given as

$$\bar{\omega}_s = \sqrt{\frac{\bar{K}_{\varepsilon\varepsilon}\bar{B}_{\phi\phi} + \bar{K}_{\phi\phi}\bar{B}_{\varepsilon\varepsilon} - \bar{K}_{\varepsilon\phi}\bar{B}_{\phi\varepsilon} - \bar{K}_{\phi\varepsilon}\bar{B}_{\varepsilon\phi}}{\bar{B}_{\varepsilon\varepsilon} + \bar{B}_{\phi\phi}}} \qquad (3.35)$$

3.2 Nonlinear Method

3.2.1 Brief Description of Trial-and-Error Method

The equations of motion $\dot{\mathbf{x}} = \mathbf{f}(\mathbf{x}, \bar{\omega})$ (Eq. 2.32 given in Chapter 2) are in a suitable form for the application of the Runge–Kutta–Fehlberg method. Consider a rotor-bearing system operating at a running speed $\bar{\omega}$ and a specified fluid viscosity μ, let the initial conditions for the nonlinear equations of motion be

$$x_1(0) = \varepsilon_0; \ x_2(0) = \dot{\varepsilon}_0; \ x_3(0) = \phi_0; \ x_4(0) = \dot{\phi}_0 \text{ when } t = 0 \qquad (3.36)$$

Starting with the initial position of (ε_0, ϕ_0) and the zero initial velocity (i.e., $\dot{\varepsilon}_0 = \dot{\phi}_0 = 0$) (Khonsari and Chang, 1993), the equations of motion ($\dot{\mathbf{x}} = \mathbf{f}(\mathbf{x}, \bar{\omega})$) can be solved using Runge-Kutta-Fehlberg method to predict the orbit of the journal center. In Runge–Kutta–Fehlberg method, the time step size is changed adaptively according to the given tolerance between the fourth- and fifth-order solutions. In the examples given in this book, the given tolerance is set to 10^{-8}.

To identify instability threshold speed, a trial-and-error method is first used to determine the initial position of the journal center for a given running speed $\bar{\omega}$ (Wang and Khonsari, 2008). The initial position of the journal center is chosen close to the steady-state equilibrium position (ε_s, ϕ_s) predicted by the equations of $\mathbf{f}(\mathbf{x}_s, \bar{\omega}) = 0$. Note that the initial position cannot coincide with the steady-state position (ε_s, ϕ_s) because, with the initial condition of (ε_s, 0, ϕ_s, 0), the system described by the equations of motion ($\dot{\mathbf{x}} = \mathbf{f}(\mathbf{x}, \bar{\omega})$) will always stay at the steady-state equilibrium position (ε_s, ϕ_s). Therefore, a small displacement perturbation is needed to identify the instability threshold (Wang and Khonsari, 2008).

The instability threshold speed is defined as the critical speed beyond which releasing the journal from a position anywhere near the steady-state equilibrium position will result in a journal orbit that tends to move away from the steady-state position and the system will become unstable (Wang and Khonsari, 2008). If the system is running at any speed less than this critical speed, releasing the journal from a position near the steady-state position will result in a journal orbit that settles onto the steady state equilibrium position. An example will be given in next section to illustrate how to identify the instability threshold speed of a rotor-bearing system with a given Sommerfeld number S using this trial-and-error method.

3.2.2 Illustration of the Trial-and-Error Method

For a long journal bearing with the commonly assumed fluid inlet condition ($\Theta_i = 0$, $\bar{p}_i = 0$) and RFJ boundary conditions (Wang and Khonsari, 2008), the

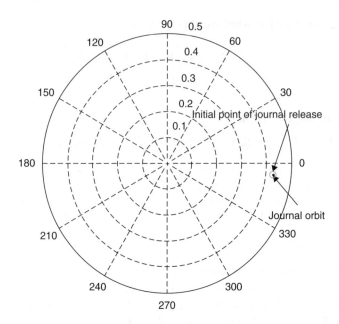

Figure 3.3   Journal orbit at $\bar{\omega} = 1.32$. From Wang and Khonsari (2008) © Elsevier Limited.

equations of motion of $\dot{\mathbf{x}} = \mathbf{f}(\mathbf{x}, \bar{\omega})$ (Eq. 2.31 given in Section 2.3) are solved directly using Runge–Kutta Fehlberg method with the initial velocity of the journal center set to zero. To predict the instability threshold speed at a derived Sommerfeld number (for example, $S = 0.04$), the trial-and-error method described in Section 3.2.1 is used to select the initial position of the journal center and the running speed $\bar{\omega}$.

Using the trial-and-error method described in Section 3.2.1, Figures 3.3, 3.4, 3.5 and 3.6 are obtained. Figures 3.4 and 3.6 are a close-up view of the journal orbit described by Figures 3.3 and 3.5, respectively. In Figures 3.3, 3.4, 3.5, and 3.6, a circle labeled "start" represents the initial release position of the journal center.

Figures 3.3 and 3.4 show the journal orbit at $\bar{\omega} = 1.32$ with initial velocity set to zero and initial position selected at $(\varepsilon, \phi) = (0.43, 84°)$, which is close to the steady-state position of the journal center at $(\varepsilon_s, \phi_s) = (0.433, 83.7°)$.

Figures 3.3 and 3.4 also show that when $\bar{\omega} = 1.32$, after releasing the journal from a position very close to the steady state equilibrium position, the journal orbit settles onto the steady-state equilibrium position. This kind of journal orbit implies that the instability threshold speed $\bar{\omega}_{st}$ should be greater than 1.32 (Wang and Khonsari, 2008).

Figures 3.5 and 3.6 show that the journal orbit at $\bar{\omega} = 1.33$ with initial velocity set to zero and initial position selected at $(\varepsilon, \phi) = (0.43, 84°)$, which is close to the steady-state equilibrium position of the journal center at $(\varepsilon_s, \phi_s) = (0.433, 83.7°)$. It is shown that when $\bar{\omega} = 1.33$, even if the shaft is released from a position which

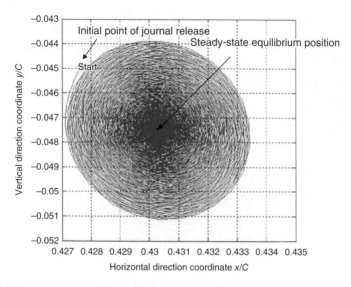

Figure 3.4   Close-up of the journal orbit shown in Figure 3.3. From Wang and Khonsari (2008) © Elsevier Limited.

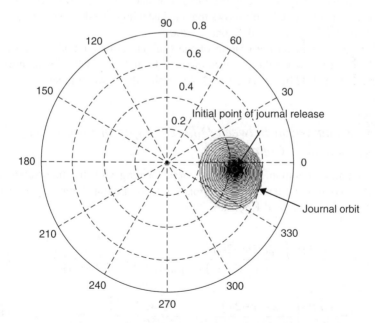

Figure 3.5   Journal orbit at $\bar{\omega} = 1.33$. From Wang and Khonsari (2008) © Elsevier Limited.

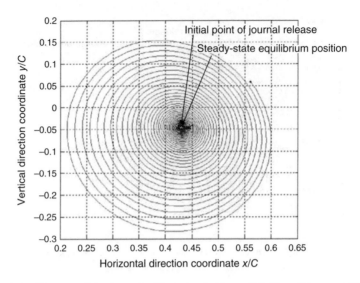

Figure 3.6 Close-up of the journal orbit shown in Figure 3.5. From Wang and Khonsari (2008) © Elsevier Limited.

is very close to the steady-state equilibrium position, the journal center will move away from the steady-state equilibrium position and the system will become unstable. This kind of journal orbit shows that the instability threshold speed $\bar{\omega}_{st}$ should be less than 1.33 (Wang and Khonsari, 2008).

Therefore, based on the criteria of the trial-and-error method, the instability threshold speed $\bar{\omega}_{st}$ for the given Sommerfeld number S of 0.04 is identified to be greater than 1.32 but less than 1.33. Averaging these two numbers, $\bar{\omega}_{st} \approx 1.325$.

3.2.3 Comparison Between Different Types of Fluid-Film Boundary Conditions

The fluid force components in long journal bearing with the half-Sommerfeld boundary condition (defined in Section 1.1.1) applied are given by Equations 3.37 and 3.38 (Wang and Khonsari, 2008).

$$f_\varepsilon = -\frac{6\mu LR^3}{C^2}\left\{\frac{2\varepsilon^2\left(\omega-2\dot{\phi}\right)}{\left(2+\varepsilon^2\right)\left(1-\varepsilon^2\right)} + \frac{\dot{\varepsilon}}{\left(1-\varepsilon^2\right)^{3/2}}\left[\pi - \frac{16}{\pi\left(2+\varepsilon^2\right)}\right]\right\} \qquad (3.37)$$

$$f_\phi = \frac{6\mu LR^3}{C^2}\left[\frac{\pi\varepsilon\left(\omega-2\dot{\phi}\right)}{\left(2+\varepsilon^2\right)\left(1-\varepsilon^2\right)^{1/2}} + \frac{4\varepsilon\dot{\varepsilon}}{\left(2+\varepsilon^2\right)\left(1-\varepsilon^2\right)}\right] \qquad (3.38)$$

Assuming fluid inlet condition of ($\Theta_i = 0$, $\bar{p}_i = 0$), Figures 3.7 and 3.8 compare the instability threshold speeds predicted based on RFJ boundary condition with those predicted based on the half-Sommerfeld boundary condition. The same equations of motion are solved while only boundary conditions, which are used to obtain the dynamic fluid force components, are different. The trial-and-error method described earlier in this section is used to obtain the instability threshold speed.

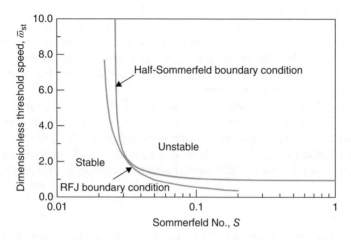

Figure 3.7 Instability threshold speeds corresponding to a series of Sommerfeld number S ($\Theta_i = 0$, $\bar{p}_i = 0$). From Wang and Khonsari (2008) © Elsevier Limited.

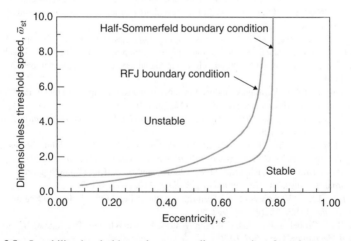

Figure 3.8 Instability threshold speed corresponding to a series of steady-state eccentricity ratio ε ($\Theta_i = 0$, $\bar{p}_i = 0$). From Wang and Khonsari (2008) © Elsevier Limited.

Figure 3.7 shows the comparison of the instability threshold speeds as a function of Sommerfeld number based on different types of boundary conditions. RFJ boundary condition (defined in chapter 1) represents the boundary conditions with consideration of an appropriate starting position of fluid cavitation, the reformation of fluid film at the end of cavitation, and the effect of fluid inlet pressure and inlet position. Figure 3.7 shows that the instability threshold speed predicted based on the assumed half-Sommerfeld boundary condition is overestimated.

Figure 3.8 compares the instability threshold speeds as a function of steady-state eccentricity ratio ε based on different types of boundary conditions. It shows that when the steady-state eccentricity ratio ε is less than 0.37, the instability threshold speed predicted based on the assumed half-Sommerfeld boundary condition is *overestimated*; when the steady-state eccentricity ratio ε is greater than 0.37, the instability threshold speed predicted based on the assumed half-Sommerfeld boundary condition is *underestimated*.

Figures 3.7 and 3.8 show that, though very low, the instability threshold speed of a lightly loaded rotor-bearing system predicted based on the half-Sommerfeld boundary condition is overestimated. These comparisons indicate that the stability of a rotor-bearing system is strongly influenced by the cavitation boundary conditions and lightly loaded journal bearings are more susceptible to instability.

References

Childs, D., Moes, H., Leeuwen, H.V., 1977, "Journal Bearing Impedance Description for Rotor Dynamic Applications," *ASME Journal of Lubrication Technology*, **99**, pp. 198–214.

Deepak, J.C., Noah, S.T., 1998, "Experimental Verification of Subcritical Whirl Bifurcation of a Rotor Supported on a Fluid Film Bearing," *ASME Journal of Tribology*, **120**, pp. 605–609.

Greenwood, D.T., 1988, *Intermediate Dynamics*, 2nd edition, Prentice Hall, Englewood Cliffs, NJ.

Holmes, R., 1960, "The Vibration of a Rigid Rotor on Short Journal Bearings," *Journal of Mechanical Engineering Science*, **2** (4), pp. 337–341.

Hori, Y., 1959, "A Theory of Oil Whip," *ASME Journal of Applied Mechanics*, **26**, pp. 189–198.

Hori, Y., 1988, "Anti-earthquake Considerations in Rotordynamics," Proceedings of the ImechE Fourth International Conference on Vibration in Rotating Machinery, Edinburgh, September 13–15, 1988, C318/88, pp. 1–8.

Hori, Y., Kato, T., 1990, "Earthquake-Induced Instability of a Rotor Supported by Oil Film Bearings," *ASME Journal of Vibration and Acoustics*, **112**, pp. 160–165.

Khonsari, M.M., Booser, E.R., 2008, *Applied Tribology: Bearing Design and Lubrication*, 2nd edition, John Wiley & Sons, Chichester.

Khonsari, M.M., Chang, Y.J., 1993, "Stability Boundary of Non-linear Orbits within Cleareance Circle of Journal Bearings," *ASME Journal of Vibration and Acoustics*, **115**, pp. 303–307.

Majumdar, B.C., Brewe, D.E., Khonsari, M.M., 1988, "Stability of a Rigid Rotor Supported on Flexible Oil Journal Bearings," *ASME Journal of Tribology*, **110**, pp. 181–187.

Newkirk, B.L., 1956, "Varieties of Shaft Disturbances due to Fluid Films in Journal Bearings," *Transactions of the ASME*, **78**, pp. 985–988.

Newkirk, B.L., Grobel, L.P., 1934, Oil-Film Whirl—A Non-Whirling Bearing, *ASME Journal of Applied Mechanics*, **1**, pp. 607–615.

Newkirk, B.L., Lewis, J.F., 1956, "Oil Film Whirl—An Investigation of Disturbances Due to Oil Films in Journal Bearings," *Transactions of the ASME*, **78**, pp. 21–27.

Newkirk, B.L., Taylor, H.D., 1925, "Shaft Whipping Due to Oil Action in Journal Bearing," *General Electric Review*, **28**, pp. 559–568.

Pinkus, O., 1953, "Note on Oil Whip," *ASME Journal of Applied Mechanics*, **75**, pp. 450–451.

Szeri, A., 1966, "Linearized Force Coefficients of A 110° Partial Journal Bearing," *Proceedings of the Institution of Mechanical Engineers*, **181** (Pt 3B), pp. 130–133.

Wang, J.K., Khonsari, M.M., 2006a, "On the Hysteresis Phenomenon Associated with Instability of Rotor-Bearing Systems," *ASME Journal of Tribology*, **128**, pp. 188–196.

Wang, J.K., Khonsari, M.M., 2006b, "Prediction of the Stability Envelope of Rotor-Bearing System," *ASME Journal of Vibration and Acoustics*, **128**, pp. 197–202.

Wang, J.K., Khonsari, M.M., 2006c, "Bifurcation Analysis of a Flexible Rotor Supported by Two Fluid Film Journal Bearings," *ASME Journal of Tribology*, **128**, pp. 594–603.

Wang, J.K., Khonsari, M.M., 2006d, "A New Derivation for Journal Bearing Stiffness and Damping Coefficients in Polar Coordinates," *Journal of Sound and Vibration*, **290**, pp. 500–507.

Wang, J.K., Khonsari, M.M., 2006e, "Application of Hopf Bifurcation Theory to the Rotor-Bearing System with Turbulent Effects," *Tribology International*, **39**, pp. 701–714.

Wang, J.K., Khonsari, M.M., 2008, "Effects of Oil Inlet Pressure and Inlet Position of Axially Grooved Infinitely Long Journal Bearings. Part II: Nonlinear Instability Analysis," *Tribology International*, **41**, pp. 132–140.

4

Introduction to Hopf Bifurcation Theory

Periodic solutions (also referred to as limit circles, periodic orbits) have been observed in many nonconservative systems, which are often governed by some nonlinear ordinary differential equations.

In 1942, Hopf was the first to point out that, subject to four important hypotheses, a nonconstant periodic orbit bifurcates at some critical system parameter. He proposed a uniqueness theorem about the bifurcation and provided information about the stability of periodic orbits. However, it was difficult to apply his theory to a specific example, especially in determining the direction of bifurcation and the stability of the periodic orbit (Myers, 1984). In 1965, Friedrichs provided criteria for distinguishing subcritical bifurcation and supercritical bifurcation (Deepak and Noah, 1998). Poore (1976) simplified the existing theories and derived algebraic criteria that can be easily applied in order to determine the direction of bifurcation and the stability of the periodic solution/orbit (Myers, 1984). Nearly 40 years after the initial publishing of the Hopf bifurcation theorem, Hassard et al. (1981) systematically developed a mature and effective Hopf bifurcation algorithm that can be applied on many autonomous systems in different fields. Hassard et al. have also demonstrated how to apply this Hopf bifurcation algorithm to some simple systems analytically and more involved systems numerically.

Thermohydrodynamic Instability in Fluid-Film Bearings, First Edition.
J. K. Wang and M. M. Khonsari.
© 2016 John Wiley & Sons, Ltd. Published 2016 by John Wiley & Sons, Ltd.

4.1 Brief Description of Hopf Bifurcation Theory

Hopf bifurcation deals with the bifurcation of the periodic orbits from the equilibrium points of a system, whose behavior is described by the ordinary differential equations $\dot{x} = f(x, \bar{\omega})$ as the system parameter $\bar{\omega}$ crosses a critical value $\bar{\omega}_{st}$. A Hopf bifurcation occurs when, as the system parameter $\bar{\omega}$ varies, a single complex conjugate pair of eigenvalues of the linearized system equations become purely imaginary in the process of crossing into the right half plane (positive real parts of eigenvalues) (Hollis and Taylor, 1986).

The Hopf bifurcation analysis is based on the following four important hypotheses (Hassard *et al.*, 1981):

  i. Equation $\dot{x} = f(x, \bar{\omega})$ has an isolated stationary point, say at $x = x_s(\bar{\omega})$;

 ii. Its Jacobian matrix $f_x(x_s(\bar{\omega}), \bar{\omega}) = \left(\dfrac{\partial f_i}{\partial x_j}(x_s(\bar{\omega}), \bar{\omega}); \, i, j = 1, \ldots, n \right)$ has exactly a pair of complex conjugate eigenvalues $\alpha(\bar{\omega}) \pm i\beta(\bar{\omega})$ such that when $\bar{\omega} = \bar{\omega}_{st}$, $\beta(\bar{\omega}_{st}) > 0$ and $\alpha(\bar{\omega}_{st}) = 0$. Meanwhile, the other $(n-2)$ eigenvalues possess purely negative real parts;

iii. f is analytic in x and $\bar{\omega}$ in a neighborhood of $(x, \bar{\omega}) = (x_s, \bar{\omega}_{st})$;

 iv. $(d\alpha(\bar{\omega})/(d\bar{\omega}))(\bar{\omega}_{st}) \neq 0$, where $\alpha(\bar{\omega})$ is the real part of the pair of eigenvalues stated above that are continuous at $\bar{\omega}_{st}$.

In the above hypotheses, $\bar{\omega}_{st}$ is called the critical value of the system parameter $\bar{\omega}$. If the assumption (ii) holds, then the assumption (iv) implies that the stability of the stationary point $x_s(\bar{\omega})$ will be lost as $\bar{\omega}$ crosses $\bar{\omega}_{st}$. Under these conditions, at the onset of bifurcation, the system has a family of periodic solutions. The Hopf bifurcation theorem provides appropriate criteria for the prediction of the existence, shape, and period of the periodic solution (Hassard *et al.*, 1981). In general, except the special degenerate case where bifurcation occurs only for $\bar{\omega} \equiv \bar{\omega}_{st}$, periodic solutions exist in only one of the two cases: either when $\bar{\omega} > \bar{\omega}_{st}$ or when $\bar{\omega} < \bar{\omega}_{st}$ (Myers, 1984). The case with periodic solutions existing when $\bar{\omega} > \bar{\omega}_{st}$ is called a supercritical bifurcation. If the periodic solutions exist only in the case of $\bar{\omega} < \bar{\omega}_{st}$, the system is said to undergo a so-called subcritical bifurcation (Wang and Khonsari, 2006).

To implement Hopf bifurcation analysis, the right-hand side of the nonlinear equations of motion $\dot{x} = f(x, \bar{\omega})$ is expanded in a Taylor series about the steady state equilibrium position $(x = x_s)$ as follows (Wang and Khonsari, 2006).

$$f(x, \bar{\omega}) = f(x_s, \bar{\omega}) + \frac{\partial f}{\partial x}(x_s, \bar{\omega}) \Delta x + \frac{\partial^2 f}{\partial x^2}(x_s, \bar{\omega})(\Delta x)^2 + \frac{\partial^3 f}{\partial x^3}(x_s, \bar{\omega})(\Delta x)^3 + \text{H.O.T.}$$

$$(4.1)$$

where $\Delta x(\bar{\omega}) = x(\bar{\omega}) - x_s(\bar{\omega})$ and H. O. T. represents the higher order terms in the Taylor series expansion. The zeroth-order terms $f(x_s, \bar{\omega})$ are used to determine the steady state equilibrium position as derived in Section 2.3. The first-order terms $\partial \mathbf{f}/\partial \mathbf{x}(\mathbf{x}_s, \bar{\omega})$—often referred to as the Jacobian matrix of the equations of motion (Eq. 4.1)—are used to determine the dynamic performance through the analysis of its eigenvalues. The second-order terms $\partial^2 \mathbf{f}/\partial \mathbf{x}^2(\mathbf{x}_s, \bar{\omega})$ and third-order terms $\partial^3 \mathbf{f}/\partial \mathbf{x}^3(\mathbf{x}_s, \bar{\omega})$ are used to determine the characteristics of the periodic solutions. Specifically, the stability, amplitude, and frequency of the periodic solutions are provided by these terms in Section 4.2.

4.2   Shape and Size and Stability of Periodic Solutions

Hassard *et al.* (1981) developed a general algorithm to calculate six bifurcation parameters using the Hopf bifurcation theory. This algorithm can be utilized directly as a part of a specific simulation package. Having evaluated the terms $\mathbf{f}_{\mathbf{x}}(\mathbf{x}_s, \bar{\omega})$ and $\partial^2 \mathbf{f}/\partial \mathbf{x}^2(\mathbf{x}_s, \bar{\omega})$ and $\partial^3 \mathbf{f}/\partial \mathbf{x}^3(\mathbf{x}_s, \bar{\omega})$, using the Hopf bifurcation formulae based on the six bifurcation parameters, the shape and size and stability of the periodic solutions close to the bifurcation point can be evaluated accordingly (Wang and Khonsari, 2006).

The periodic solutions (also referred to as system's response) can be defined by the following Hopf bifurcation formulae (Hassard *et al.*, 1981).

$$\mathbf{x}(t, \bar{\omega}) = \mathbf{x}_s(\bar{\omega}_{\text{st}}) + \left(\frac{\bar{\omega} - \bar{\omega}_{\text{st}}}{\gamma_2}\right)^{1/2} \text{Re}\left(e^{2\pi i t / T} \mathbf{v}_1\right) + O(\bar{\omega} - \bar{\omega}_{\text{st}}) \qquad (4.2)$$

$$T(\bar{\omega}) = \frac{2\pi}{\beta(\bar{\omega}_{\text{st}})} \left[1 + \tau_2 \left(\frac{\bar{\omega} - \bar{\omega}_{\text{st}}}{\gamma_2}\right) + O(\bar{\omega} - \bar{\omega}_{\text{st}})^2\right] \qquad (4.3)$$

$$S_p(\bar{\omega}) = \beta_2 \left(\frac{\bar{\omega} - \bar{\omega}_{\text{st}}}{\gamma_2}\right) + O(\bar{\omega} - \bar{\omega}_{\text{st}})^2 \qquad (4.4)$$

where $T(\bar{\omega})$ is the period of the periodic solution at $\bar{\omega}$ and $S_p(\bar{\omega})$ is the characteristic exponent that determines its orbital stability. Equations 4.2–4.4 involve six bifurcation parameters ($\bar{\omega}_{\text{st}}$, γ_2, $\mathbf{v}_1$, $\beta(\bar{\omega}_{\text{st}})$, τ_2, and β_2) for a specific nonconservative system. γ_2 represents the parameter that governs the range of the existence of periodic solutions. If $\gamma_2 < 0$, periodic solutions exist when $\bar{\omega} < \bar{\omega}_{\text{st}}$ and subcritical bifurcation exists; if $\gamma_2 > 0$, periodic solutions exist when $\bar{\omega} > \bar{\omega}_{\text{st}}$ and supercritical bifurcation exists. A system with $\gamma_2 = 0$ is degenerate and need not be pursued further since bifurcation occurs only when $\omega \equiv \omega_{\text{st}}$. τ_2 is the expansion coefficient of the periods of periodic solutions as the system parameter $\bar{\omega}$ moves away from the critical value $\bar{\omega}_{\text{st}}$. β_2 is the leading coefficient in the expansion of a characteristic exponent $S_p(\bar{\omega})$. If $S_p(\bar{\omega}) < 0$ (i.e., $\beta_2 < 0$), the periodic solution is orbital-asymptotically stable with an asymptotic phase (see Section 4.3 for definition). Any trajectory with an initial value in the vicinity of the periodic solution tends to the periodic solution as time goes to infinity. If $S_p(\bar{\omega}) > 0$ (i.e., $\beta_2 > 0$), the periodic solution is orbital-asymptotically unstable and any trajectory with an initial value in the vicinity of the periodic solution tends to move away from the periodic solution as time goes to infinity. The vector parameter $\mathbf{v}_1$ is an eigenvector of the Jacobian matrix at the stationary point when $\bar{\omega} = \bar{\omega}_{st}$. $\mathbf{v}_1$ is normalized so that its first non-vanishing element is 1 (Hassard et al., 1981).

Equations 4.2–4.4 show that if supercritical bifurcation occurs when the system parameter $\bar{\omega}$ crosses the critical value $\bar{\omega}_{st}$, the amplitude of the stable periodic solution will ramp up. If subcritical bifurcation occurs when the system parameter $\bar{\omega}$ is crossing the critical value $\bar{\omega}_{st}$, the amplitude of the unstable periodic solution shrinks to a point at $\bar{\omega}_{st}$.

4.3 Definition of Orbital-Asymptotically Stable with an Asymptotic Phase

Let $p(t)$ be a periodic solution to $\dot{\mathbf{x}} = \mathbf{f}(\mathbf{x})$, where $\mathbf{f}$ is analytic in $\mathbf{x}$ (Hassard et al., 1981). Then $p(t)$ is asymptotically, orbitally stable with the asymptotic phase if and only if there exists an $\varepsilon' > 0$ such that if $\psi(t)$ is any solution to $\dot{\mathbf{x}} = \mathbf{f}(\mathbf{x})$ for which $|\psi(t_0) - p(t)| < \varepsilon'$ at the initial time t_0, then there exists a constant ϕ', called the asymptotic phase, with the property of

$$\lim_{t \to \infty} |\psi(t) - p(t + \phi')| = 0 \qquad (4.5)$$

Chapter 5 will show how this HBT can be applied to a rotor bearing system to study the system instability associated with fluid-film journal bearings.

References

Deepak, J.C., Noah, S.T., 1998, "Experimental Verification of Subcritical Whirl Bifurcation of a Rotor Supported on a Fluid Film Bearing," *ASME Journal of Tribology*, **120**, pp. 605–609.

Friedrichs, K.O., 1965, *Advanced Ordinary Differential Equations*, Gordon and Breach, New York.

Hassard, B.D., Kazarinoff, N.D., Wan, Y.H., 1981, *Theory and Applications of Hopf Bifurcation*, London Mathematical Society Lecture Notes 41, Cambridge University Press, New York.

Hopf, E., 1942, "Abzweigung einer Periodischen Longsung von einer Stationoren Losung einer Differential-Systems," *Berichten der Mathematisch-Physischen Klasse der Sächsischen Akademie der Wissenschaften zu Leipzig*, **95**, pp. 3–22.

Hollis, P., Taylor, D.L., 1986, "Hopf Bifurcation to Limit Cycles in Fluid Film Bearings," *ASME Journal of Tribology*, **108**, pp. 184–189.

Myers, C.J., 1984, "Bifurcation Theory Applied to Oil Whirl in Plain Cylindrical Journal Bearings," *ASME Journal of Applied Mechanics*, **51**, pp. 244–250.

Poore, A.B., 1976, "On the Theory and Application of the Hopf-Friedrichs Bifurcation Theory," *Archive for Rational Mechanics and Analysis*, **55**, pp. 61–90.

Wang, J.K., Khonsari, M.M., 2006, "Application of Hopf Bifurcation Theory to the Rotor-Bearing System with Turbulent Effects," *Tribology International*, **39**, pp. 701–714.

5

Application of HBT to Fluid-Film Bearings

This chapter focuses on how the Hopf bifurcation theory (HBT) presented in Chapter 4 can be applied to explain different phenomena associated with the stability of fluid-film journal bearings. Some of the relevant publications are as follows. The first application of HBT for the instability analysis of a rigid rotor symmetrically supported by two infinitely long journal bearings was presented by Myers in 1984. Two important phenomena—subcritical bifurcation and supercritical bifurcation—were introduced in connection with the instability of a rotor bearing system. A similar analysis, but using the short bearing theory, was published by Hollis and Taylor (1986). Sundararajan (1996) and Noah (1995) analyzed a more general case of finite journal bearings using HBT and discussed the effects of different length/diameter ratios on the subcritical bifurcation and supercritical bifurcation regimes. Deepak and Noah (1998) provided experimental evidence that verified the existence of the subcritical bifurcation of a single disk rotor supported on a short journal bearing. Wang and Khonsari (2006a, 2006b) extended the application of HBT to turbulent flow and flexible rotor. These methods have also been successfully applied to explain important phenomena such as the stability envelope, hysteresis, and the dip phenomenon influence of oil inlet temperature on the instability threshold speed of rotor bearing systems (Wang and Khonsari, 2006c, 2006d). In addition, using HBT, the influence of drag force on the dynamic performance of rotor bearing systems has been studied extensively by Wang and Khonsari (2005). These concepts are described in detail in Chapter 6.

Thermohydrodynamic Instability in Fluid-Film Bearings, First Edition.
J. K. Wang and M. M. Khonsari.
© 2016 John Wiley & Sons, Ltd. Published 2016 by John Wiley & Sons, Ltd.

In this chapter, as an introduction to the applications of HBT to the instability analysis of fluid-film journal bearings, two important phenomena of stability envelope and hysteresis will be introduced and explained using HBT.

5.1   Application I: Prediction of Stability Envelope

We begin this section by presenting a method for predicting the stability envelope of a rotor bearing system using the Hopf bifurcation theory (HBT). Results are compared with those obtained using a trial-and-error method reported by Khonsari and Chang (1993).

5.1.1   Definition of Stability Envelope

In 1993, Khonsari and Chang introduced the concept of the stability envelope, which offers an insight into the behavior of bearing instability, particularly as related to the initial conditions imposed on the system. According to their paper, there is a distinct region of stability within a closed boundary (called stability envelope R_s) that encircles the steady-state equilibrium position. Figure 5.1 shows a stability envelope R_s in bearing clearance circle. If a journal is released from inside the R_s (such as point A in Figure 5.1a) with zero initial velocity, its orbit will settle onto the steady-state equilibrium position. If released from outside the R_s (such as point B in Figure 5.1b) with zero initial velocity, the orbit will grow larger until the system reaches the so-called whip condition where the orbit extends to the clearance circle of the rotor bearing system, endangering the system's operation (Wang and Khonsari, 2006c). The shape and size of the R_s depend on the specifications of the rotor bearing system.

The concept of R_s provides a means for explaining the shock effects on the occurrence of the oil whip discovered in a classical paper by Newkirk and Taylor (1925). They pointed out that some rotor bearing systems that were running quietly at speeds lower than the threshold speed ω_{st} began to whip violently just due a "special shock." Sometimes, even a slight shock was sufficient to initiate violent whipping. Hori and Kato confirmed this phenomenon in 1959. They did extensive research on how an earthquake—which can be regarded as a shock—affects the occurrence of the oil whip of a rotor bearing system (Hori, 1988; Hori and Kato, 1990). The "special shock" in Newkirk and Taylor's paper (1925) can be considered as a perturbation whose magnitude and relative position is located outside the R_s. The magnitude of the shock, or the earthquake, which can cause the quietly running rotor bearing system to whip, is defined by the shape and size of the R_s.

The determination of the shape and size of the stability envelope R_s is important in predicting whether an external perturbation can cause a system to lose stability even when its operating speed is well below the instability threshold speed. In Khonsari and Chang's paper (1993), a trial-and-error method was proposed to determine R_s corresponding to a set of system operating parameters. For a given

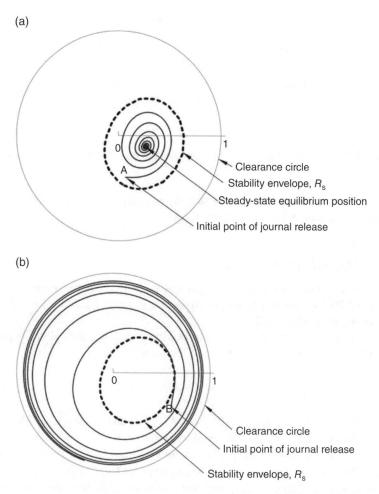

(a)

Clearance circle

Stability envelope, R_s

Steady-state equilibrium position

Initial point of journal release

(b)

Clearance circle

Initial point of journal release

Stability envelope, R_s

Figure 5.1   Stability envelope R_s in the clearance circle. (a) Stable if the journal is released inside the stability envelope R_s. (b) Unstable if the journal is released outside the stability envelope R_s. From Wang and Khonsari (2006c) © ASME.

attitude angle, bisection method (also called binary search method) was used to determine the point on the R_s. According to the authors, this trial-and-error method is "straightforward but requires rather lengthy computations." An exhaustive study using this method is, therefore, not practical, and the development of an alternative approach is desirable.

A powerful approach for predicting the stability envelope R_s is introduced by Wang and Khonsari (2006c), which is based on the Hopf bifurcation theory (HBT). To illustrate the method, a rotor bearing system consisting of a rigid rotor symmetrically supported by two identical fluid-film journal bearings is analyzed.

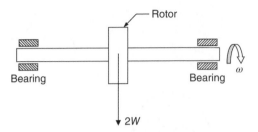

Figure 5.2 Rigid rotor supported by two journal bearings (Wang and Khonsari, 2006c)

The basic governing equations in the stability analysis using HBT are briefly summarized in Section 5.1.3 for completeness.

5.1.2 Equations of Motion

Consider a system consisting of a rigid and perfectly balanced rotor (with the weight of 2W) symmetrically supported by two identical fluid-film journal bearings as shown in Figure 5.2.

The equations of motion for this rotor bearing system are derived in Chapter 2. Written in polar coordinates, they are of the form

$$\dot{\mathbf{x}} = \mathbf{f}(\mathbf{x}, \bar{\omega}) \tag{5.1}$$

where $\mathbf{x} = (x_1, x_2, x_3, x_4)$, $x_1 = \varepsilon$, $x_2 = \dot{\varepsilon}$, $x_3 = \phi$, $x_4 = \dot{\phi}$, and $\bar{\omega} = \omega\sqrt{C/g}$. The equations of motion (Eq. 5.1) are only valid when $0 < \varepsilon < 1$.

The steady-state equilibrium position $\mathbf{x}_s$ in terms of $(x_{1s} = \varepsilon_s, x_{2s} = \dot{\varepsilon}_s, x_{3s} = \phi_s, x_{4s} = \dot{\phi}_s)$ can be easily recast in the required form for Hopf bifurcation anaysis—that is, $\mathbf{x}_s = (\varepsilon_s, 0, \phi_s, 0)$ with the governing equations of the form $\mathbf{f}(\mathbf{x}_s, \bar{\omega}) = 0$.

For simplicity, let us apply the short bearing theory with the half-Sommerfeld boundary conditions (see Chapter 1 for details) to the Reynolds equation. The short bearing theory lends itself to analytical formulation for the hydrodynamic bearing forces so that the expanded form of Equation 5.1 reads as follows (see Section 2.2.1 for details):

$$\dot{x}_1 = x_2$$

$$\dot{x}_2 = x_1 x_4^2 - \frac{\Gamma}{\bar{\omega}} \left[\frac{2x_1^2(1 - 2x_4)}{\left(1 - x_1^2\right)^2} + \frac{\pi\left(1 + 2x_1^2\right)x_2}{\left(1 - x_1^2\right)^{5/2}} \right] + \frac{1}{\bar{\omega}^2}\cos x_3$$

$$\dot{x}_3 = x_4$$

$$\dot{x}_4 = -\frac{2x_2 x_4}{x_1} + \frac{\Gamma}{\bar{\omega}} \left[\frac{\pi(1 - 2x_4)}{2\left(1 - x_1^2\right)^{3/2}} + \frac{4x_2}{\left(1 - x_1^2\right)^2} \right] - \frac{1}{\bar{\omega}^2 x_1}\sin x_3$$

$$\tag{5.2}$$

where g is the gravitational constant and Γ is a dimensionless system characteristic number $\Gamma = \mu R L^3 / \left(2mC^{2.5} g^{0.5} \right)$.

The steady state equilibium position $\mathbf{x}_s$ of the system is given by

$$\frac{x_{1s}\sqrt{16x_{1s}^2 + \pi^2\left(1 - x_{1s}^2\right)}}{\left(1 - x_{1s}^2\right)^2} = \frac{1}{S\pi(L/D)^2}$$

$$x_{2s} = 0 \qquad\qquad\qquad (5.3)$$

$$x_{3s} = \tan^{-1}\left(\frac{\pi\sqrt{1 - x_{1s}^2}}{4x_{1s}}\right)$$

$$x_{4s} = 0$$

5.1.3   Application of Hopf Bifurcation Theory to the Equations of Motion

The equations of motion (Eq. 5.1) possess a steady state equlibrium position $\mathbf{x}_s$ (the stationary point, (Eq. 5.3)) and satisfy the four hypotheses (given in Section 4.1) for the application of the Hopf bifurcation theory (HBT). Here, the system running speed $\bar{\omega}$ is the system parameter in HBT. This applies when all the other parameters of the rotor bearing system including oil viscosity are fixed. According to HBT, if the system parameter $\bar{\omega}$ in the equations of motion (Eq. 5.1) becomes greater than the critical value $\bar{\omega}_{st}$, an isolated stationary point $\mathbf{x}_s(\bar{\omega})$ loses its linear stability because the real part of a complex conjugate pair of its eigenvalues becomes non-negative. Hence, this critical value $\bar{\omega}_{st}$ is actually the threshold speed of the instability (Wang and Khonsari, 2006c).

To apply the Hopf bifurcation theory, first, the right-hand side of the decomposed nonlinear equations of motion (Eq. 5.1) is expanded in a Taylor series about the steady state equilibrium position $(\mathbf{x} = \mathbf{x}_s)$ as follows (Wang and Khonsari, 2006c):

$$\mathbf{f}(\mathbf{x},\bar{\omega}) = \mathbf{f}(\mathbf{x}_s,\bar{\omega}) + \frac{\partial\mathbf{f}}{\partial\mathbf{x}}(\mathbf{x}_s,\bar{\omega})\Delta\mathbf{x} + \frac{\partial^2\mathbf{f}}{\partial\mathbf{x}^2}(\mathbf{x}_s,\bar{\omega})(\Delta\mathbf{x})^2 + \frac{\partial^3\mathbf{f}}{\partial\mathbf{x}^3}(\mathbf{x}_s,\bar{\omega})(\Delta\mathbf{x})^3 + \text{H.O.T.}$$

$$(5.4)$$

where $\Delta\mathbf{x}(\bar{\omega}) = \mathbf{x}(\bar{\omega}) - \mathbf{x}_s(\bar{\omega})$ and H. O. T. represents the higher order terms in the Taylor series expansion. The zeroth order $\mathbf{f}(\mathbf{x}_s,\bar{\omega})$ is used to determine the steady state equilibrium position. The first order $\partial\mathbf{f}/\partial\mathbf{x}(\mathbf{x}_s,\bar{\omega})$—usually referred to as Jacobian matrix—is used to determine the dynamic performance through the analysis of its eigenvalues. The second order $\partial^2\mathbf{f}/\partial\mathbf{x}^2(\mathbf{x}_s,\bar{\omega})$ and third order $\partial^3\mathbf{f}/\partial\mathbf{x}^3(\mathbf{x}_s,\bar{\omega})$ are used to determine the characteristics of the periodic solutions. Specifically, the existence, stability, amplitude, and frequency of the periodic solutions are provided by these two terms (Hassard et al., 1981).

The periodic solutions (also referred to as the system responses) of the journal orbit are governed by the following Hopf bifurcation formulae when the shaft running speed $\bar{\omega}$ is close to the critical value $\bar{\omega}_{st}$ (Hassard *et al.*, 1981).

$$\mathbf{x}(t,\bar{\omega}) = \mathbf{x}_s(\bar{\omega}_{st}) + \left(\frac{\bar{\omega}-\bar{\omega}_{st}}{\gamma_2}\right)^{1/2} \mathrm{Re}\left(e^{2\pi it/T}\mathbf{v}_1\right) + O(\bar{\omega}-\bar{\omega}_{st}) \qquad (5.5)$$

$$T(\bar{\omega}) = \frac{2\pi}{\beta(\bar{\omega}_{st})}\left[1 + \tau_2\left(\frac{\bar{\omega}-\bar{\omega}_{st}}{\gamma_2}\right) + O(\bar{\omega}-\bar{\omega}_{st})^2\right] \qquad (5.6)$$

$$S_p(\bar{\omega}) = \beta_2\left(\frac{\bar{\omega}-\bar{\omega}_{st}}{\gamma_2}\right) + O(\bar{\omega}-\bar{\omega}_{st})^2 \qquad (5.7)$$

where $T(\bar{\omega})$ represents the period of the periodic solution of the journal orbit at speed $\bar{\omega}$ and $S_p(\bar{\omega})$ is the characteristic exponent that determines its orbital stability. For a specific rotor bearing system with a given oil viscosity μ, Equations 5.5–5.7 involve six bifurcation parameters ($\bar{\omega}_{st}$, γ_2, τ_2, β_2, $\beta(\bar{\omega}_{st})$, and $\mathbf{v}_1$). The existence range of the periodic solutions of the journal orbit is governed by the parameter γ_2. τ_2 is the coefficient in the expansion of the periods of periodic solutions as the running speed $\bar{\omega}$ either increases (when $\gamma_2 > 0$) or decreases (when $\gamma_2 < 0$) away from the critical value $\bar{\omega}_{st}$. Equations 5.5–5.7 are not valid when $\gamma_2 = 0$. β_2 is the leading coefficient in the expansion of the characteristic exponent. Vector parameter $\mathbf{v}_1$ is an eigenvector of the Jacobian matrix at the stationary point when $\bar{\omega} = \bar{\omega}_{st}$.

For a specific system with the predicted six bifurcation parameters, Table 5.1 shows the criteria about the range of the existence of the periodic solutions, the bifurcation type (either subcritical or supercritical), and whether the periodic solutions are stable or not.

Table 5.1 also reveals that the value of $((\bar{\omega}-\bar{\omega}_{st})/\gamma_2)$ is always positive. Therefore, the characteristic exponent $S_p(\bar{\omega})$ given by Equation 5.7 always possesses the same sign as the bifurcation parameter β_2.

According to the stability definition of the periodic solutions discussed in Chapter 4, when the periodic solution is asymptotically stable, any trajectory with an initial value in the neighborhood of the periodic solution tends to the periodic

Table 5.1 Bifurcation characteristics (Wang and Khonsari, 2006c)

γ_2	$\gamma_2 < 0$	Periodic solutions exist for $\bar{\omega} < \bar{\omega}_{st}$; subcritical bifurcation exists
	$\gamma_2 > 0$	Periodic solutions exist for $\bar{\omega} > \bar{\omega}_{st}$; supercritical bifurcation exists
β_2	$\beta_2 < 0$	Periodic solutions are orbital-asymptotically stable
	$\beta_2 > 0$	Periodic solutions are orbital-asymptotically unstable

solution as time goes to infinity. On the other hand, if the periodic solution is asymptotically unstable, any trajectory with an initial value in the neighborhood of the periodic solution tends to move away from the periodic solution as time goes to infinity. The definition of the unstable periodic solution directly corresponds to that of the stability envelope R_s introduced in Section 5.1.1 (Wang and Khonsari, 2006c). It correlates that if the journal is released from a position inside the R_s, the system tends to asymptotically approach the steady-state equilibrium position; if the journal is released from a position outside the R_s, the journal orbit of the system tends to asymptotically approach the clearance circle, i.e., becomes unstable. Therefore, HBT provides a direct method to predict the unstable periodic solution, i.e., the R_s.

The interpretation of the stable periodic solution is straightforward and the effect of any perturbation on the system stability decays quickly. Concerning the unstable periodic solution, the implication is that a sufficiently large perturbation may cause a stable system to lose stability. In other words, the stability of the system is dependent upon the magnitude of the initial perturbation, as originally hypothesized by Khonsari and Chang (1993).

According to the definitions of oil whirl and oil whip in Chapter 3, oil whirl is a tendency of a journal to whirl moderately with a stable trajectory while oil whip is a violent whipping motion triggered by a perturbation with its amplitude located outbound of the stability envelope. It is straightforward to correlate the oil whirl with supercritical bifurcation and the oil whip with subcritical bifurcation. If supercritical bifurcation occurs when the running speed of the rotor bearing system crosses the threshold speed, the journal will start whirling at the stable periodic solution and the amplitude of the oil whirl or stable periodic solution will ramp up as running speed changes. If subcritical bifurcation occurs when the running speed of the rotor bearing system is approaching the threshold speed, a sufficiently large perturbation will cause the journal to suddenly lose its stability and whip. In other words, the amplitude of the oil whip will jump up to such a large orbit that the journal tends to asymptotically approach the clearance circle and could rub with the bearing surface. The oil whip can be triggered by a perturbation whose amplitude depends on the proximity of the running speed to the threshold speed. The closer to the threshold speed, the less is the required perturbation amplitude to trigger the oil whip. At and beyond the threshold speed, no perturbation is required and the system whips inherently.

5.1.4   Numerical Investigation of the Stability Envelope R_s

The system of equations (Eq. 5.1) is also in a suitable form for a numerical solution using the Runge–Kutta–Fehlberg method. Consider a rotor-bearing system operating at a running speed $\bar{\omega}$ and a given oil viscosity μ; let the initial conditions for the equations of motion be

$$x_1(0) = \varepsilon_0; \; x_2(0) = \dot{\varepsilon}_0; \; x_3(0) = \phi_0; \; x_4 = \dot{\phi}_0 \quad \text{when } t = 0 \qquad (5.8)$$

The selection of the initial position of the journal can play an important role in determining the orbital stability. Following the trial-and-error method introduced in Section 3.2, the nonlinear equations of motion (Eq. 5.1) are solved using the Runge–Kutta–Fehlberg method while the initial velocity $(\dot{\varepsilon}_0, \dot{\phi}_0)$ is set to zero. In the Runge–Kutta–Fehlberg method, the time step size is changed adaptively according to a given tolerance between the fourth- and fifth-order solutions. In the simulations reported in this section, the given tolerance is set to 10^{-8}. To determine R_s, for a given attitude angle, bisection method is used to determine the point on the R_s. After repeating the same procedure for all possible attitude angles, the stability envelope R_s is obtained by connecting all the predicted points sequentially.

5.1.5 Illustrative Case Study

Consider a rotor bearing system whose specifications are listed in Table 5.2. This rotor bearing system consists of a rigid rotor symmetrically supported by two identical fluid-film journal bearings.

Let us apply the HBT as described in Section 5.1.3 as well as the trial-and-error numerical method presented in Sections 5.1.4 and 3.2 to determine the stability envelope R_s of this rotor bearing system.

5.1.5.1 Stability Envelope Predicted using HBT

Based on the equations presented in Section 5.1.3, the Hopf bifurcation parameters are obtained and the results are shown in Table 5.3.

Table 5.2 Specification of a rotor bearing system (Wang and Khonsari, 2006c)

Journal diameter, D	0.0254 m
Length of the bearing, L	0.0127 m
Radial clearance, C	50.8×10^{-6} m
Mass of the rotor, $2m$	5.4523 kg
Oil viscosity, μ	0.01 Pa·s

Table 5.3 Hopf bifurcation parameters for the system described in Table 5.2 (Wang and Khonsari, 2006c)

$\bar{\omega}_{st}$	γ_2	τ_2	β_2	$\beta(\bar{\omega}_{st})$	v_1	$x_s(\omega_{st})$
2.64	−1.7540	0.3962	0.3088	0.5151	1.0 + 0.0i	0.2502
					0.0 + 0.5151i	0
					0.5801 − 4.7457i	1.2528
					2.4444 + 0.2988i	0

Table 5.3 shows that the dimensionless threshold speed $\bar{\omega}_{st} = 2.64$, where $\bar{\omega}_{st} = \omega_{st}\sqrt{C/g}$. Since $\gamma_2 < 0$ and $\beta_2 > 0$, according to Table 5.1, the bifurcation is subcritical and periodic solutions exist only when $\bar{\omega} < \bar{\omega}_{st} = 2.64$ and they are unstable. Therefore, the stability envelope R_s exists for system running speeds that are close to but less than the threshold speed $\bar{\omega}_{st}$.

With the obtained Hopf bifurcation parameters in Table 5.3, the periodic solutions of the system are fully defined by the Hopf bifurcation formulae (5.5–5.7) described in Section 5.1.3. Figure 5.3 shows the predicted periodic solutions of the rotor bearing system described in Table 5.2.

Figure 5.3a shows the shapes of the periodic solutions of journal orbit as a function of dimensionless running speed that is close to but below the critical speed $\bar{\omega}_{st} = 2.64$. Each cross section of Figure 5.3a at a given dimensionless running speed $\bar{\omega}$ is the periodic solution of the journal orbit at this running speed. Figure 5.3a shows that the periodic solutions of journal orbit shrink to a single point as the running speed approaches the critical value, $\bar{\omega}_{st}$. Figure 5.3b shows the bifurcation profile, which depicts the amplitude of the periodic solution as a function of system running speed close to but below the critical speed $\bar{\omega}_{st}$. The subcritical bifurcation profile shrinks to a single point as the running speed approaches the critical value, $\bar{\omega}_{st}$. The amplitude of the periodic solution of the journal orbit corresponding to a specific running speed $\bar{\omega}$ is symmetrical at ε_s and is bounded by $\varepsilon_s \pm \sqrt{(\bar{\omega}-\bar{\omega}_{st})/\gamma_2}$. Figure 5.3c, as an example, shows the unstable periodic solution, R_s, when the dimensionless running speed $\bar{\omega} = 2.63$. According to the definition of the R_s, Figure 5.3c shows that, at $\bar{\omega} = 2.63$, if the journal is released inside the R_s, the orbit of the journal of the system asymptotically approaches the steady-state equilibrium position; if the journal is released outside the R_s, the journal orbit tends to asymptotically approach the clearance circle, that is, becomes unstable.

5.1.5.2   Stability Envelope Using Trial-and-Error Method

To illustrate the concept, we examine the stability envolpe corresponding to three different dimensionless running speeds $\bar{\omega}$ of 2.63, 2.60, and 2.55. All three cases correspond to a speed below the dimensionless threshold speed of $\bar{\omega}_{st} = 2.64$. Three stability envelopes R_s predicted by HBT will be compared with those determined by the trial-and-error method given in Section 5.1.4. The results are shown in Figures 5.4, 5.5, and 5.6.

In Figures 5.4, 5.5, and 5.6, the solid line represents the R_s predicted using HBT. The discrete dots represent the R_s obtained by the trial-and-error method. These figures show that the shapes of these two stability envelopes are similar. The R_s determined by HBT predicts a slightly smaller area than that obtained by the trial-and-error method. If the journal is released from inside the R_s (such as Point A), the orbit of the journal settles onto a steady-state point position. If released from

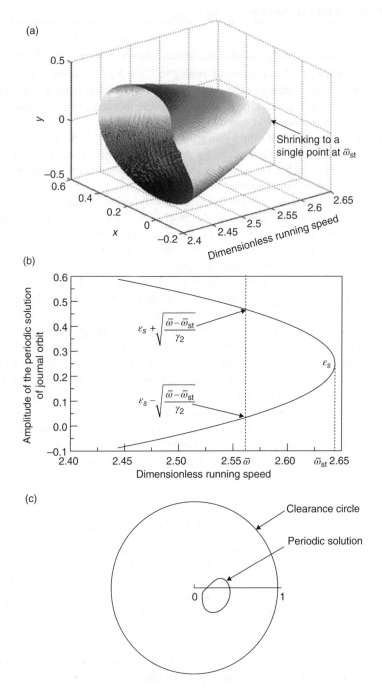

Figure 5.3 The periodic solutions of the equations of motion. (a) Periodic solutions in x and y coordinates in the unit clearance circle. (b) Subcritical bifurcation profile. (c) Periodic solution when $\bar{\omega} = 2.63$. From Wang and Khonsari (2006c) © ASME.

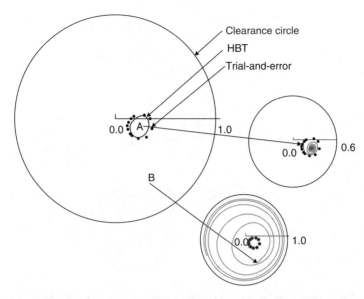

Figure 5.4 Comparison of the two stability envelopes R_s at $\bar{\omega} = 2.63$. From Wang and Khonsari (2006c) © ASME.

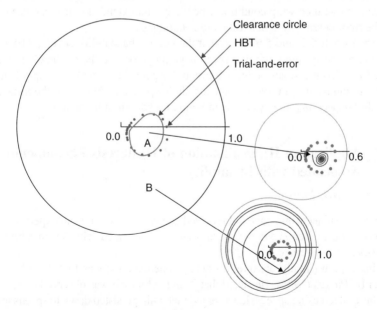

Figure 5.5 Comparison of the two stability envelopes R_s at $\bar{\omega} = 2.60$. From Wang and Khonsari (2006c) © ASME.

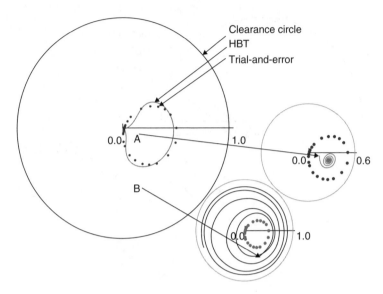

Figure 5.6   Comparison of the two stability envelopes R_s at $\bar{\omega} = 2.55$. From Wang and Khonsari (2006c) © ASME.

outside the R_s (such as Point B), the orbit grows larger and larger until the system reaches a so-called whip condition where the orbit extends to the clearance circle of the rotor bearing system and failure is imminent.

Figures 5.4, 5.5, and 5.6 reveal that the size of the stability envelope becomes smaller as the system's running speed is getting closer to the predicted threshold speed $\bar{\omega}_{st}$. Even a small perturbation could trigger the system to whip when the system running speed is close to the threshold speed $\bar{\omega}_{st}$. At or after the instability threshold speed $\bar{\omega}_{st}$, the system is in whipping condition by default.

5.2   Application II: Explanation of Hysteresis Phenomenon Associated with Instability

5.2.1   Introduction

One of the subjects that have not received adequate attention is an experimentally observed phenomenon known as "hysteresis" associated with the instability of rotor-bearing system.

This experimentally observed hysteresis phenomenon was first described in a paper by Pinkus (1956). He stated that "when whipping was observed under conditions of decreasing speeds, it was noted that whip persisted down to speeds lower than those at which whip started when the speed was being increased." In 1959, Hori confirmed the occurrence of this phenomenon. However, this phenomenon did not recieve much attention until Hori presented his findings in the IMechE Fourth International Conference on Vibration in Rotating Machinery in 1988. Subsequently, Guo and Adams (1995, 1996) published some trial-and-error simulations and experimental results and tried to explain the nature of the hysteresis phenomenon. They summed up that an "elusive unstable intermediate solution" exists in the hysteresis loop. Horattas (1996) and Horattas et al. (1997) reported the results of a set of experiments that confirmed the existence of the hyteresis phenomenon and verified some of the results reported by Guo and Adams. Muszynska (1998) proposed a qualitative explanation of the hysteresis phenomenon from the viewpoint of circumferential velocity ratios. According to Muszynska, in the runup process, the fluid circumferential average velocity ratio (especially that of the fluid damping force), which is driven by the rotor rotation, is lagging behind (i.e., smaller than) that in the rundown process. Yet, interestingly, a quantitative explanation of the hysteresis phenomenon had remained unresolved due to the lack of an appropriate analytical approach until Wang and Khonsari (2006d) successfully predicted the existence and profile of the hysteresis phenomenon using the Hopf bifurcation theory (HBT).

In this section, we present an analytical approach to predict the hysteresis phenomenon using HBT. We show that the existence and the profile of hysteresis phenomenon are dependent upon the system's operating parameters. To this end, the effect of oil viscosity on the hysteresis phenomenon and its implications on the rotor bearing system instability are described.

5.2.2 Definition of Hysteresis Phenomenon Associated with Instability

In performing oil whirl/whip instability experiments, the threshold of instability is normally detected by gradually increasing the system operating speed (hereinafter, referred to as the run-up process) and monitoring the dynamic behavior of the system. After crossing the threshold of instability and upon gradually decreasing the system running speed (hereinafter, referred to as the run-down process), one often observes that the oil whip disappears at a running speed that is below the threshold speed detected during the run-up process. This hysteresis phenomenon is, in fact, repeatable and occurs even though all other system parameters including oil viscosity remain unchanged. The characteristics of the hysteresis phenomenon associated with a perfectly balanced, lightly loaded rotor bearing system without any misalignment are illustrated in Figure 5.7. The speed at which the oil whip starts is called the run-up threshold speed (RUTS); the speed at which the oil whip disappears is called the rundown threshold speed (RDTS) (Wang and Khonsari, 2006d).

Referring to Figure 5.7, we start by increasing the system running speed from the stable state where the perturbation is nil until the evidence of instability is detected at Point A. Based on the definitions in Chapter 3 and the discussions in Section 5.1,

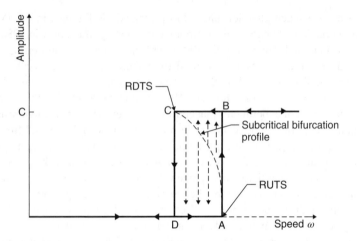

Figure 5.7   Illustration of the hysteresis phenomenon. From Wang and Khonsari (2006d) © ASME.

the type of instability (sudden amplitude jump instead of ramping up) is oil whip. The speed at which it occurs is identified as RUTS. At this speed the vibration amplitude suddenly jumps to Point B. At this stage, the inherent perturbation is large and the system is unstable. Now, upon reducing speed, we note that at Point C, the oil whip suddenly disappears and the vibration amplitude shrinks to Point D (a steady state position). Point C is identified as the RDTS. Between RUTS and RDTS, there is a curve called subcritical bifurcation profile (upper symmetrical half). The definition and characteristics of subcritical bifurcation profile are given in Section 5.1.

The stability of the rotor bearing system for any running speed between RUTS and RDTS depends on how large the perturbation is (Wang and Khonsari, 2006d). If the amplitude of the perturbation is located inside the subcritical bifurcation profile, the system will return to the steady-state position. If the amplitude of the perturbation is outside the subcritical bifurcation profile, the perturbation will trigger oil whip and the system will become unstable. Physically, this is analogous to the stability envelope discussed in Section 5.1. So, when the releasing point of the journal is outside the unstable periodic solution (i.e., the R_s), the system exhibits orbitally unstable behavior. If the amplitude of the perturbation is inside the subcritical bifurcation profile, analogous to releasing the journal inside the stability envelope (i.e., the R_s), the vibration caused by the perturbation will decay quickly and the system will be stable.

The hysteresis phenomenon occurs only in the case of subcritical bifurcation while it does not exist at all in the case of supercritical bifurcation because only subcritical bifurcation cases have unstable periodic solutions (i.e., stability envelopes). The subcritical bifurcation profile determines the profile of the hysteresis loop as shown in Figure 5.7.

5.2.3   Experimental Investigation

Having discussed the nature of the hysteresis phenomenon and the associated RUTS and RDTS, we now present the results of some typical experiments to gain further insight (Wang and Khonsari, 2006d). Figure 5.8 shows the schematic of a lightly loaded rotor symmetrically supported by two identical hydrodynamic journal bearings at two ends. The specifications of this rotor bearing system are listed in Table 5.4. This test rig is equipped with a heating and cooling system (not shown) capable of supplying oil with controllable inlet oil temperature from 0 to 180°C.

The centrally loaded rotor is driven by a half-horsepower AC–DC series motor with an electronic speed controller through a flexible spring coupling. The running speed range is 0–10 000 rpm. In the preliminary trial tests, the system is found to be always stable when the running speed is below 7500 rpm. To measure the temperature distribution of the bearing pad, eight thermocouples are embedded inside one of the journal bearings. Vibration data (journal orbits) are recorded in real time by a pair of XY eddy current proximity probes located inside the journal bearing housing. The vibration signals are processed through a computerized data acquisition system. The data is analyzed using ADRE® (Automated Diagnostics for Rotating

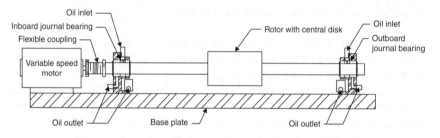

Figure 5.8   Schematic figure of the test rig (Chauvin, 2003).

Table 5.4   Specification of the rotor bearing system (Wang and Khonsari, 2006d)

Journal diameter, D	25.4 mm
Length of journal bearing, L	12.7 mm
Span length between two bearings, l	527.1 mm
Outer diameter of middle weight	76.2 mm
Length of middle weight	127.0 mm
Inside diameter of the hollow shaft, d	15.2 mm
Mass of the rotor, $2m$	5.4523 kg
Radial clearance, C	0.0508 mm
Lubricant grade	ISO 32
Inlet pressure, p_{in}	31 kPa
Inlet temperature range, T_{in}	0–180°C

Equipment), a commercially available software. The Bode plots and waterfall plots presented in this section are generated using this software.

The test procedure for the hysteresis profile is summarized as follows (Wang and Khonsari, 2006d).

1. To reach the desired inlet oil temperature, oil is heated/cooled and circulated through the whole system at a low speed such as 1800 rpm until the inlet oil temperature and the bearing temperatures are stabilized.
2. The system running speed is then increased up to 7000 rpm at an increment of about 200 rpm.
3. Starting from 7000 rpm, the system running speed increment is decreased to about 50 rpm. To determine the instability threshold speed accurately, the speed is ramped up as slowly as possible when approaching the instability threshold speed, which is determined during the preliminary trial tests.
4. After crossing the threshold speed of instability, the system running speed is increased further to a higher value to get a sustaining oil whip with its amplitude (peak-to-peak) maintained at about two times the radial clearance C.
5. Then, the system running speed is decreased as slowly as possible until the oil whip disappears.
6. At last, the system is allowed to coast down.

Note that every time the system running speed is changed, the system is allowed to run long enough to reach a steady state, which is identified as a constant temperature reading of the bearing bushing.

The results of two different but typical cases will be shown to illustrate the nature of the hysteresis phenomenon associated with instability (Wang and Khonsari, 2006d). One deals with a high inlet oil temperature and the other with a low temperature. As we shall see, the lubricant temperature plays a significant role.

Case I: Observation of Hysteresis Phenomenon at High Inlet Oil Temperature
In the test, oil inlet temperature is set at 53°C. At this temperature the viscosity of ISO 32 oil is 16.0 mPa·s (1 mPa·s = 0.001 Pa·s).

Figure 5.9 is the Bode plot of the overall vibration amplitude (peak to peak). Figure 5.10 is the waterfall plot corresponding to Figure 5.9. The Bode plot shows how the overall vibration amplitude changes along with the system running speed. The waterfall plot shows the frequency and amplitude of the subsynchronous vibration component as well as those of the synchronous vibration component at each system running speed recorded. In these plots, the amplitude and frequency of the vibration are shown in mm (millimeter) and Hz (Hertz) while the system running speed is shown in rpm (revolutions per minute).

The Bode plot shown in Figure 5.9 illustrates that, upon slowly increasing the system running speed (i.e., run-up process), oil whip instability is detected at about 7980 rpm (RUTS); upon slowly reducing the speed (i.e., run-down process), the oil whip disappears at about 7760 rpm (RDTS). This plot shows an experimentally

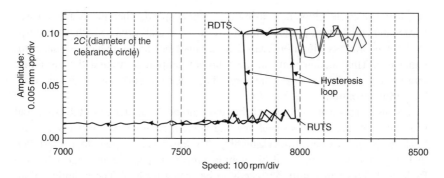

Figure 5.9 Bode plot showing hysteresis phenomenon at high oil inlet temperature ($T = 53°$C). From Wang and Khonsari (2006d) © ASME.

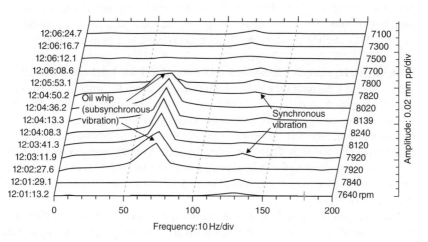

Figure 5.10 The waterfall plot corresponding to Figure 5.9. From Wang and Khonsari (2006d) © ASME.

obtained hysteresis-like behavior, illustrating a clear difference between RUTS and RDTS. Note that the differences between RDTS and RUTS will increase if the system running speed is changed rapidly. But, even if the running speed of the rotor is changed at a very small rate, a clear hysteresis phenomenon still exists. In addition, during a series of experiments devoted to this investigation, intermittent noise indicative of the rubbing between the journal and the bushing was audiable when the subsynchronous vibration (oil whip) started and this effect persisted. This rubbing noise also proves that the journal orbit is very close to the clearance circle when the oil whip starts.

Another interesting observation is that, upon increasing the running speed and then crossing RUTS, the vibration amplitude suddenly jumps up close to the clearance circle. Upon decreasing the running speed, when the system crosses RDTS, the vibration amplitude suddenly resumes a small value, which is attributed to the synchronous vibration caused by a small but unavoidable residual rotor unbalance. This sudden jump in the vibration amplitude around the instability threshold is one of the important features of the subcritical bifurcation and its unstable periodic solutions (i.e., the R_s).

It is also important to point out that the vibration amplitudes in the Bode and waterfall plots are a peak-to-peak reading while the amplitudes of periodic solutions are given as a zero-to-peak reading. To draw the hysteresis loop illustrated in Figure 5.7 with subcritical bifurcation profile (upper symmetrical half), the Bode plot shown in Figure 5.9 needs to be scaled by half. The bounding limit of the Bode plot is scaled down from $2C$ (the diameter of clearance circle) to C (the radius of clearance circle).

In addition, the nominal radial clearance C is specified as 0.0508 mm. However, due to the manufacturing tolerances, the actual radial clearance is slightly greater than 0.0508 mm. Therefore, the diameter of the clearance circle, as shown in Figure 5.9, is slightly greater than 0.102 mm (Wang and Khonsari, 2006d).

Case II: No Hysteresis Phenomenon Observed at Low Inlet Oil Viscosity
Let us review the vibration results of another case when the inlet oil viscosity is 7.0 mPa·s, which corresponds to the inlet oil temperature at 80°C for ISO 32 oil. All the other system parameters are identical to those in Case I. In this case, preliminary trial tests have shown that the system is also stable when the running speed is below 7500 rpm. Therefore, only the portion of the vibration data with the system running speed higher than 7500 rpm will be analyzed. Figure 5.11 presents the Bode plot of the overall vibration. Figure 5.12 is the waterfall plot corresponding to Figure 5.11.

The Bode plot in Figure 5.11 reveals that in this case there is no distinct difference between the run-up process and the run-down process, implying that the hysteresis

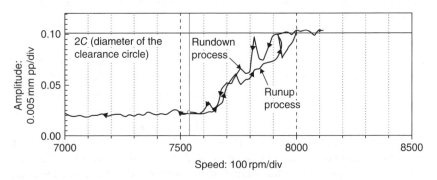

Figure 5.11 No hysteresis phenomenon is observed at low inlet oil temperature ($T = 80°C$). From Wang and Khonsari (2006d) © ASME.

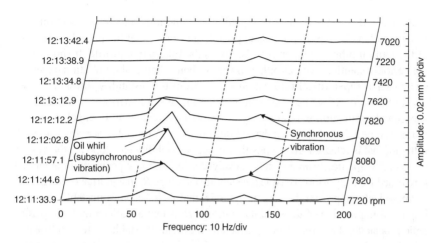

Figure 5.12   The waterfall plot corresponding to Figure 5.11. From Wang and Khonsari (2006d) © ASME.

phenomenon does not exist. Here, the amplitude of oil whirl gradually ramps up after the system running speed crosses the instability threshold speed and gradually runs up. Upon decreasing the system running speed, the amplitude of oil whirl gradually decreases from near the clearance circle to a very small value. This gradual change of the vibration amplitude above while close to the instability threshold speed is one of the important features of the supercritical bifurcation and its stable periodic solutions.

Figures 5.9, 5.10, 5.11, and 5.12 also show that the amplitude of the synchronous vibration component in Figures 5.9 and 5.10 is less than that in Figures 5.11 and 5.12 when the system's running speed is less than RDTS. As pointed out by Bhat *et al.* (1982), the oil viscosity has a significant influence on the unbalance response (synchronous vibration) of a rotor supported by two fluid-film journal bearings. Generally speaking, decreasing the oil viscosity (i.e., increasing the oil temperature) while maintaining all other system parameters unchanged will escalate the unbalance response of the rotor. Therefore, the amplitude of the synchronous vibration (unbalance response) in Figure 5.11 is higher than that in Figure 5.9 since the oil viscosity in the experiment whose results are shown in Figure 5.11 is less than that of the experiment whose results are shown in Figure 5.9 (Wang and Khonsari, 2006d).

5.2.4   Relationship between Hysteresis Phenomenon and Subcritical Bifurcation

Let us now apply the Hopf bifurcation theory (HBT) to predict the subcritical bifurcation profile and to help gain further insight into the hysteresis phenomenon observed in Section 5.2.3.

Recall HBT analysis of the equations of motion derived in Section 5.1. Equations 5.5–5.7 describe the size, period, and stability of the periodic solutions of the journal orbit. Equation 5.5 also gives the bifurcation profile. It shows how the periodic solution evolves either toward (when subcritical bifurcation exists) or away (when supercritical bifurcation exists) from the instability threshold speed $\bar{\omega}_{st}$. In Section 5.1, we have also shown that the unstable periodic solution when subcritical bifurcation exists directly corresponds to the stability envelope R_s.

Let us repeat the definition of the stability envelope R_s (unstable periodic solution): if the journal is released from a position inside the stability envelope R_s, the system tends to asymptotically approach the steady-state equilibrium position; if the journal is released from a position outside the stability envelope R_s, the journal orbit tends to asymptotically approach the clearance circle, that is, it becomes unstable. Another important implication of the stability envelope R_s is that a sufficiently large perturbation located outside the stability envelope may cause a stable system to lose stability. A sudden shock, such as an earthquake, could trigger an oil whip in a rotor bearing system that was running stably. The perturbation amplitude capable of transferring a stable journal into a violent whipping condition depends on the proximity of the system running speed to the instability threshold speed. The closer to the instability threshold speed, the less is the perturbation amplitude required to trigger oil whip. Figure 5.3a in Section 5.1 shows that the unstable periodic solution (i.e., stability envelope) shrinks to a "single point" as the running speed approaches the instability threshold speed, $\bar{\omega}_{st}$. At and after $\bar{\omega}_{st}$, an oil whip exists even in the absence of any perturbation.

Therefore, the definition of the stability envelope R_s (i.e., unstable periodic solution) provides information about the circumstance when the oil whip starts or disappears (Wang and Khonsari, 2006c). The stability envelope R_s exists only when the bifurcation of a rotor bearing system is subcritical. The subcritical bifurcation profile depicts how the amplitude of the stability envelope R_s (i.e., unstable periodic solution) changes with the system running speed, which is close to but less than the instability threshold speed $\bar{\omega}_{st}$. As an example, Figure 5.3b in Section 5.1 shows a subcritical bifurcation profile. It also shows that the subcritical bifurcation profile shrinks to a single point as the system running speed approaches the instability threshold speed, $\bar{\omega}_{st}$.

With the subcritical bifurcation profile and the definition of the stability envelope R_s (i.e., unstable periodic solution), the hysteresis loop described in Figure 5.7 can be explained (Wang and Khonwsari, 2006d). As the system running speed is increased from the stable state, where the perturbation is nil or has a small amplitude such as synchronous vibration due to residual rotor unbalance, an occurrence of oil whip would be first detected at RUTS (point A), where the vibration amplitude suddenly jumps to point B since the small amplitude of the perturbation crosses and moves outside of the subcritical bifurcation profile. However, upon decreasing the running speed from the unstable condition—where the perturbation is very large—to a lower speed (i.e., at RDTS, point C), the oil whip disappears and the vibration amplitude suddenly shrinks to point D since the amplitude of the perturbation is crossing and moving inside of the subcritical bifurcation profile. Thus, the subcritical bifurcation profile defines the hysteresis phenomenon.

It is important to note that there always exists some residual rotor unbalance in a rotor bearing system and a synchronous whirl of the shaft journal due to the residual unbalance is always present. Relative to the ideal steady-state equilibrium position, the synchronous whirl can be treated as a perturbation. Therefore, in reality, the run-up threshold speed is always less than the ideal instability threshold speed $\bar{\omega}_{st}$.

5.2.5 Case Studies

Referring to Section 5.1.3, the dimensionless system running speed $\bar{\omega}$ ($\bar{\omega} = \omega\sqrt{C/g}$) is the system parameter for the application of the Hopf bifurcation theory (HBT). For the rotor bearing system described in Table 5.4, we apply the HBT to predict the bifurcation profile at a given viscosity while maintaining all the other system parameters fixed (Wang and Khonsari, 2006d). Table 5.5 shows the predicted results of the six parameters $\bar{\omega}_{st}$, γ_2, $\mathbf{v}_1$, $\beta(\bar{\omega}_{st})$, τ_2, β_2 and $\mathbf{x}_s(\bar{\omega}_{st})$ subject to a change in oil viscosity in the range from $3.0\,\text{mPa·s}$ to $30.0\,\text{mPa·s}$.

Table 5.5   Six bifurcation parameters (Wang and Khonsari, 2006d)

	Oil viscosity (mPa·s)						
	30.0	10.0	7.5	7.1	7.0	5.0	3.0
$\bar{\omega}_{st}$	2.7449	2.6436	2.5971	2.5884	2.5862	2.5432	2.5475
γ_2	−20.7492	−1.7540	−0.2558	−0.0043	0.0600	1.6289	5.8289
τ_2	0.1380	0.3962	0.5309	0.5611	0.5692	0.8071	1.6289
β_2	2.3281	0.3088	0.0437	0.0007	−0.0101	−0.2469	−0.6715
$\beta(\bar{\omega}_{st})$	0.5024	0.5152	0.5203	0.5212	0.5214	0.5238	0.5124
$\mathbf{v}_1$	1.0 + 0.0i	1.0 +	1.0 +	1.0 +	1.0 +	1.0 +	1.0 +
	0.0 +	0.0i	0.0i	0.0i	0.0i	0.0i	0.0i
	0.50i	0.0 +	0.0 + 0.	0.0 +	0.0 +	0.0 +	0.0 +
	0.63 −	0.51i	52i	0.52i	0.52i	0.52i	0.51i
	11.19i	0.58 −	0.53 −	0.52 −	0.52 −	0.43 −	0.17 −
	5.62 +	4.74i	4.18i	4.10i	4.09i	3.85i	3.98i
	0.31i	2.44 +	2.17 +	2.14 +	2.13 +	2.01 +	2.04 +
		0.29i	0.28i	0.27i	0.27i	0.22i	0.09i
$\mathbf{x}_s(\bar{\omega}_{st})$	0.0914	0.2502	0.3118	0.3241	0.3273	0.4033	0.5100
	0	0	0	0	0	0	0
	1.4544	1.2529	1.1750	1.1595	1.1555	1.0594	0.9241
	0	0	0	0	0	0	0

(a)

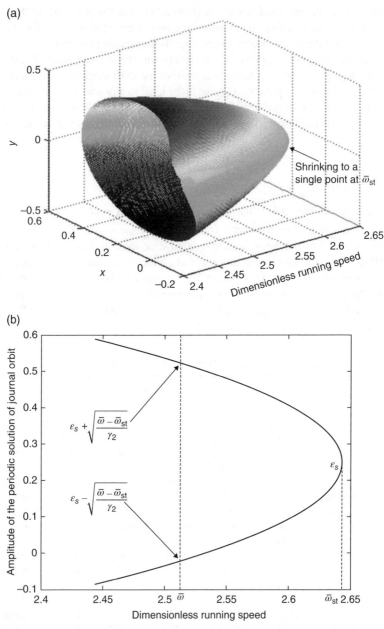

(b)

Figure 5.13 Periodic solutions when oil viscosity is 10.0 mPa·s. (a) Bifurcation profile—3D. (b) Bifurcation profile—2D. (c) Periodic solution. From Wang and Khonsari (2006d) © ASME.

(c)

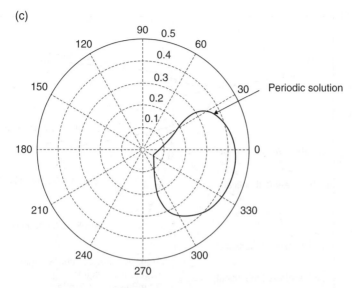

Figure 5.13 (Continued)

Based on these six bifurcation parameters, the periodic solutions can be predicted by Equations 5.5–5.7. For example, if the oil viscosity is equal to 10.0 mPa·s, the dimensionless threshold speed $\bar{\omega}_{st}$ is 2.6436, where $\bar{\omega}_{st} = \omega_{st}\sqrt{C/g}$. Since $\gamma_2 < 0$ and $\beta_2 > 0$, unstable periodic solutions exist for $\bar{\omega} \leq \bar{\omega}_{st} = 2.6436$ and the bifurcation is subcritical. Figure 5.13a–c is plotted based on the six parameters corresponding to the case where the oil viscosity is equal to 10.0 mPa·s using the Hopf bifurcation formulae (Eqs. 5.5–5.7) given in Section 5.1.3.

Figure 5.13a shows the shapes of the periodic solutions of journal orbit as a function of system running speed that is close to but less than or equal to the threshold speed $\bar{\omega}_{st} = 2.6436$. Each cross section of Figure 5.13a at a given dimensionless running speed $\bar{\omega}$ is a periodic solution of the journal orbit. Figure 5.13c, for example, shows the periodic solution with the dimensionless running speed $\bar{\omega} = 2.63$. Figure 5.13a also shows that the periodic solutions of journal orbit shrink to a single point as the running speed approaches the threshold speed $\bar{\omega}_{st}$.

Figure 5.13b shows the subcritical bifurcation profile, which depicts the amplitudes of the periodic solutions as a function of system running speed close to but less than or equal to the threshold speed $\bar{\omega}_{st}$. The amplitude of the periodic solution at a specific running speed $\bar{\omega}$ is bounded by $\varepsilon_s \pm \sqrt{(\bar{\omega}-\bar{\omega}_{st})/\gamma_2}$. In addition, Figure 5.13b shows that the bifurcation profile shrinks to a single point as the system running speed approaches the threshold speed $\bar{\omega}_{st}$.

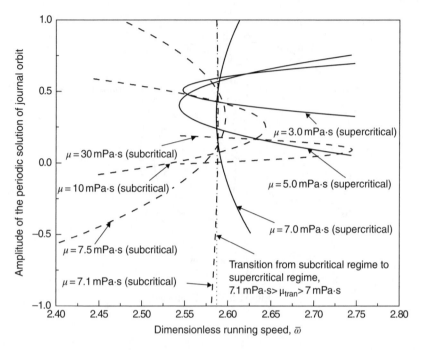

Figure 5.14   Bifurcation profiles at different viscosities. From Wang and Khonsari (2006d) © ASME.

Using the Hopf bifurcation Equations 5.5–5.7 presented in Section 5.1.3, similar results as shown in Figure 5.13b can be generated for the other different oil viscosities listed in Table 5.5. Figure 5.14 shows the bifurcation profiles at these different oil viscosities.

Examination of Figure 5.14 and Table 5.5 shows that the amplitude of the periodic solution is a function of the deviation of the system operating speed $\bar{\omega}$ from the instability threshold speed $\bar{\omega}_{st}$ and the oil viscosity. For the oil viscosity μ ranging from 7.1 mPa·s up to 30 mPa·s, since $\gamma_2 < 0$ and $\beta_2 > 0$, unstable periodic solutions associated with subcritical bifurcation exist for $\bar{\omega} < \bar{\omega}_{st}$. For the oil viscosity μ ranging from 7.0 mPa·s down to 3.0 mPa·s, since $\gamma_2 > 0$ and $\beta_2 < 0$, stable periodic solutions associated with supercritical bifurcation exist for $\bar{\omega} > \bar{\omega}_{st}$. The transition viscosity μ_{tran} from subcritical bifurcation to supercritical bifurcation is between 7.0 mPa·s and 7.1 mPa·s.

The bifurcation profiles presented in Figure 5.14 have the following features in terms of oil viscosity μ and shaft running speed $\bar{\omega}$. In every subcritical bifurcation case with the oil viscosity higher than μ_{tran}, the amplitude of the *unstable* periodic solution will *decrease* as the running speed increases. In every supercritical bifurcation case with the oil viscosity lower than μ_{tran}, the amplitude of the *stable* periodic solution will *increase* up to the clearance circle as the running speed increases.

After obtaining the bifurcation profile corresponding to a specific set of system parameters, the existence and characteristics of hysteresis phenomenon can be easily predicted based on the correlation developed in Section 5.2.4. If the bifurcation type is subcritical, hysteresis phenomenon exists and the hysteresis loop is determined by the subcritical bifurcation profile. If the bifurcation type is supercritical, the hysteresis phenomenon does not exist at all.

From the above analysis, it can be concluded that by changing the inlet oil temperature and even oil grade if necessary, one can adjust the oil viscosity to change the bifurcation type of rotor bearing systems.

In light of the above discussion, the results of the two typical experimental cases presented in Section 5.2.3 can now be interpreted. The only system operating parameter that changes from Case I to Case II is the inlet oil temperature. The different phenomena shown in Figures 5.9 and 5.11 are caused by different oil viscosities. In Case I where $\mu = 16.0$ mPa·s, the system undergoes subcritical bifurcation and a distinct hysteresis phenomenon is observed. However, in Case II where $\mu = 7.0$ mPa·s, the system undergoes supercritical bifurcation and the hysteresis phenomenon does not exist at all. These results also attest to the importance of thermal effects on the stability of hydrodynamic bearings. It is important to point out that the oil viscosity referred in the experimental results is at the inlet oil temperature while the oil viscosity referred in all simulation results is the effective oil viscosity of the oil film. In Chapter 6, we show further evidence of the importance of thermal effects, particularly associated with the change in the oil inlet temperature. Accurate determination of the hysteresis profile requires a complete thermohydrodynamic analysis (THD) of a fluid-film bearing to assess the temperature and then viscosity field within the fluid film (Khonsari and Beaman, 1986; Khonsari, 1987; Khonsari *et al.*, 1996; Fillon and Khonsari, 1996; Keogh *et al.*, 1997 and 2001; Jang and Khonsari, 2004; Singhal and Khonsari, 2005). This is beyond the scope of the present section.

In addition, Table 5.5 shows that the predicted threshold speeds $\bar{\omega}_{st}$ are much higher than the experimental ones such as the RUTS (around 1.9 in dimensionless form) in Figure 5.9. One of the major reasons for this discrepancy is that the Hopf bifurcation analysis in this section is based on a rigid-rotor model where the actual rotor stiffness is not taken into account. Generally speaking, rotor flexibility tends to lower the predicted threshold speed significantly (Wang and Khonsari, 2006b). Chapter 6 shows how rotor flexibility affects the instability threshold speed of rotor bearing systems.

The limited number of examples given in this section also show that, for the specific rotor bearing system described in Table 5.4, with oil viscosity in the range from 3.0 mPa·s to 30.0 mPa·s, there is a single transition oil viscosity μ_{tran} (about 7.1 mPa·s). As oil viscosity decreases across μ_{tran}, the type of bifurcation changes from subcritical to supercritical and the hysteresis phenomenon disappears.

In fact, more extensive and fundamental research (Wang and Khonsari, 2006b) has shown that the type of Hopf bifurcation solely depends on the system characteristic number $\Gamma = \mu R L^3 / \left(2mC^{2.5} g^{0.5} \right)$ and rotor stiffness $\bar{K} = k(C/mg)$. For a specific rotor bearing system with a low or moderate rotor stiffness, there are generally two transition system characteristic numbers (Γ_1 and Γ_2, $\Gamma_1 < \Gamma_2$). Subcritical bifurcation and hysteresis phenomenon exist for both higher than Γ_2 and lower than Γ_1 system characteristic numbers. The Hopf bifurcation is supercritical for the rotor bearing system with intermediate system characteristic numbers ($\Gamma_1 < \Gamma < \Gamma_2$). Section 6.3 is devoted to this topic.

Since the definition of the system characteristic number $\Gamma = \mu R L^3 / \left(2mC^{2.5} g^{0.5} \right)$ reveals that Γ is a linear function of oil viscosity μ, for a given rotor bearing system, Γ can be easily controlled by changing the oil viscosity μ. In other words, for a specific rotor bearing system with a low or moderate rotor stiffness, there are two transition oil viscosities μ_1 and μ_2 ($\mu_1 < \mu_2$). Subcritical bifurcation and hysteresis phenomenon exist for any oil viscosity which is higher than μ_2 or lower than μ_1. To avoid the hysteresis phenomenon and oil whip, the oil viscosity must be maintained within the range from μ_1 up to μ_2.

References

Adams, M.L., Guo, J.S., 1996, "Simulations and Experiments of the Non-linear Hysteresis Loop for Rotor-Bearing Instability," IMechE Conference Transactions 1996-6, Sixth International Conference on Vibration in Rotating Machinery, September 9-12, Mechanical Engineering Publication Ltd. London, C500/001/96, pp. 309–319.

Bhat, R.B., Rao, J.S., Sankar, T.S., 1982, "Optimum Journal Bearing Parameters for Minimum Rotor Unbalance Response in Synchronous Whirl," *ASME Journal of Mechanical Design*, **104**, pp. 339–344.

Chauvin, D., 2003, "An Experimental Investigation of Whirl Instability Including Effects of Lubricant Temperature in Plain Circular Journal Bearings," MS Thesis, Department of Mechanical Engineering, Louisiana State University, Baton Rouge, LA.

Deepak, J.C., Noah, S.T., 1998, "Experimental Verification of Subcritical Whirl Bifurcation of a Rotor Supported on a Fluid Film Bearing," *ASME Journal of Tribology*, **120**, pp. 605–609.

Fillon, M., Khonsari, M.M., 1996, "Thermohydrodynamic Design Charts for Tilting-Pad Journal Bearings," *ASME Journal of Tribology*, **118**, pp. 232–238.

Guo, J.S., Adams, M.L., 1995, "Characteristics of the Nonlinear Hysteresis Loop for Rotor-Bearing Instability," DE-Vol. 84-2, *Proceedings of 1995 ASME Design Engineering Technical Conferences*, **3**(Part B), pp. 1081–1091.

Hassard, B.D., Kazarinoff, N.D., Wan, Y.H., 1981, *Theory and Applications of Hopf Bifurcation*, London Mathematical Society Lecture Notes 41, Cambridge University Press, New York.

Hollis, P., Taylor, D.L., 1986, "Hopf Bifurcation to Limit Cycles in Fluid Film Bearings," *ASME Journal of Tribology*, **108**, pp. 184–189.

Horattas, G.A., 1996, "Experimental Investigation of Dynamic Nonlinearities in Rotating Machinery," Dissertation, Department of Mechanical and Aerospace Engineering, Case Western Reserve University, Cleveland, OH.

Horattas, G.A., Adams, M.L., Abdel Magied, M.F., and Loparo, K.A., 1997, "Experimental Investigation of Dynamic Nonlinearities in Rotating Machinery," Proceedings of the ASME Design Engineering Technical Conferences, September 14–17, Sacramento, CA, pp. 1–12.

Hori, Y., 1959, "A Theory of Oil Whip," *ASME Journal of Applied Mechanics*, **26**, pp. 189–198.

Hori, Y., 1988, "Anti-earthquake Considerations in Rotordynamics," Proceedings of the ImechE Fourth International Conference on Vibration in Rotating Machinery, Edinburgh, September 13-15, Mechanical Engineering Publication Ltd., C318/88, pp. 1–8.

Hori, Y., Kato, T., 1990, "Earthquake-Induced Instability of a Rotor Supported by Oil Film Bearings," *ASME Journal of Vibration and Acoustics*, **112**, pp. 160–165.

Jang, J.Y., Khonsari M.M., 2004, "Design of Bearings Based on Thermohydrodynamic Analysis," *Journal of Engineering Tribology, Proceedings of Institution of Mechanical Engineers, Part J*, **218**, pp. 355–363.

Keogh, P., Gomiciaga, R., Khonsari, M.M., 1997, "CFD Based Design Techniques for Thermal Prediction in a Generic Two-axial Groove Hydrodynamic Journal Bearing," *ASME Journal of Tribology*, **119**, pp. 428–436.

Keogh, P., Khonsari, M.M., 2001, "Influence of Inlet Conditions on the Thermohydrodynamic State of a Fully circumferentially Grooved Journal Bearing," *ASME Journal of Tribology*, **123**, pp. 525–532.

Khonsari, M.M., Beaman, J., 1986, "Thermohydrodynamic Analysis of Laminar Incompressible Journal Bearings," *ASLE Transactions*, **29**, pp. 141–150.

Khonsari, M.M., 1987, "A Review of Thermal Effects in Hydrodynamic Bearings. Part II: Journal Bearings," *ASLE Transactions*, **30**(1), pp. 26–33.

Khonsari, M.M., Chang, Y.J., 1993, "Stability Boundary of Non-linear Orbits within Cleareance e of Journal Bearings," *ASME Journal of Vibration and Acoustics*, **115**, pp. 303–307.

Khonsari, M., Jang, J., Fillon, M., 1996, "On the Generalization of Thermohydrodynamic Analysis for Journal Bearings," *ASME Journal of Tribology*, **118**, pp. 571–579.

Muszynska, A., 1998, "Transition to Fluid-Induced Limit Cycle Self-Excited Vibrations of a Rotor and Instability Threshold 'Hysteresis'," Proceedings of ISROMAC-7, The Seventh International Symposium on Transport Phenomena and Dynamics of Rotating Machinery, February 22-26, Bird Rock Publication House, Honolulu, HI, pp. 775–784.

Myers, C.J., 1984, "Bifurcation Theory Applied to Oil Whirl in Plain Cylindrical Journal Bearings," *ASME Journal of Applied Mechanics*, **51**, pp. 244–250.

Newkirk, B.L., Taylor, H.D., 1925, "Shaft Whipping Due to Oil Action in Journal Bearing," *General Electric Review*, **28**, pp. 559–568.

Noah, S.T., 1995, "Significance of Considering Nonlinear Effects in Predicting the Dynamic Behavior of Rotating Machinery," *Journal of Vibration and Control*, **1**, pp. 431–458.

Pinkus, O., 1956, "Experimental Investigation of Resonant Whip," *Transactions of the ASME*, **78**, pp. 975–983.

Singhal, S., Khonsari, M.M., 2005, "A Simplified Thermohydrodynamic Stability Analysis of Journal Bearings," *Journal of Engineering Tribology, Proceedings of Institution of Mechanical Engineers, Part J*, **229**, pp. 225–234.

Sundararajan, P., 1996, "Response and Stability of Nonlinear Rotor Bearing Systems," Dissertation, Department of Mechanical Engineering, Texas A&M University, College Station, TX.

Wang, J.K., Khonsari, M.M., 2005, "Influence of Drag Force on the Dynamic Performance of Rotor-Bearing System," *Proceedings of the Institution of Mechanical Engineers, Part J, Journal of Engineering Tribology*, **219**(4), pp. 291–295.

Wang, J.K., Khonsari, M.M., 2006a, "Application of Hopf Bifurcation Theory to the Rotor-Bearing System with Turbulent Effects," *Tribology International*, **39**(7), pp. 701–714.

Wang, J.K., Khonsari, M.M., 2006b, "Bifurcation Analysis of a Flexible Rotor Supported by Two Fluid Film Journal Bearings," *ASME Journal of Tribology*, **128**, pp. 594–603.

Wang, J.K., Khonsari, M.M., 2006c, "Prediction of the Stability Envelope of Rotor-Bearing System," *ASME Journal of Vibration and Acoustics*, **128**, pp. 197–202.

Wang, J.K., Khonsari, M.M., 2006d, "On the Hysteresis Phenomenon Associated with Instability of Rotor-Bearing Systems," *ASME Journal of Tribology*, **128**, pp. 188–196.

6

Analysis of Thermohydrodynamic Instability

In Chapters 3 and 5, conventional and HBT-based (based on Hopf bifurcation theory) instability analyses were applied to a hypothetically rigid rotor supported by two identical plain journal bearings. In reality, rotors possess some degree of flexibility, and imperfections such as residual shaft unbalance and bearing bushing wear always exist. In this chapter, we address how these factors affect the stability of a rotor-bearing system. In addition, the operating conditions of a rotor-bearing system directly influence the system stability (Wang and Khonsari, 2006d, 2008b). Aside from the running speed and load, the operating conditions that need to be specified properly often include the fluid inlet temperature, inlet pressure, and inlet position. The effects of these influencing factors on the instability of the rotor-bearing system are also discussed in this chapter. Based on the analyses, general design guidelines will be recommended at the end of the section. We end this chapter by presenting the effects of turbulent flow and drag force on the system thermohydrodynamic instability.

6.1 Inlet Temperature Effects

The literatures published before 2006 contain a number of disparities in assessing the effect of inlet fluid temperature on the instability threshold speed of a rotor-bearing system associated with fluid-film bearings. A brief discussion of the relevant papers follows.

Thermohydrodynamic Instability in Fluid-Film Bearings, First Edition.
J. K. Wang and M. M. Khonsari.
© 2016 John Wiley & Sons, Ltd. Published 2016 by John Wiley & Sons, Ltd.

Maki and Ezzat (1980) are probably the first to clearly discuss the importance of the thermal effects on bearing instability and the associated disparities in the literature. They pointed out that the influence of inlet oil temperature on bearing instability reported by Newkirk and Lewis (1956) was opposite to that reported by Pinkus (1953, 1956). Specifically, Newkirk and Lewis concluded that the instability threshold speed increased along with the increasing of the inlet oil temperature. However, Pinkus reported the results of several experiments where a journal bearing whipped at high inlet oil temperatures did not whip at all with cold inlet oil. In addition, he pointed out that "raising the oil temperature tends, in general, to lower the speed at which whip starts." In essence, Pinkus' experimental results (1953) implied that a lower inlet oil temperature had a stabilizing effect on rotor-bearing systems. His conclusion is contrary to that of Newkirk and Lewis (1956).

Wang and Khonsari (2006b, 2006d) carried out a series of experimental investigations devoted to the understanding of the effect of inlet fluid temperature on the instability threshold of a rotor-bearing system in 2006. Their results revealed the existence of a dip in the instability threshold speed plotted as a function of the fluid inlet temperature. As fluid inlet temperature increases, the instability threshold speed first slowly decreases, then reaches a valley, and subsequently increases at a steep rate. This "dip phenomenon"—which was shown to be persistent and reproducible in all the experiments—can be used to explain the 'peculiar' inconsistencies described by Newkirk (1956): "In the field, some rotors exhibiting the disturbance can be quieted by warming up the oil supplied to the bearing and in other cases, cooling the oil supply is effective."

6.1.1   Theoretical Prediction

Before presenting the experimental results on the inlet fluid temperature effects, it is appropriate to provide some background theoretical predictions based on a simplified model. Figure 6.1 shows a planar model for a centrally loaded flexible rotor supported on two fluid-film journal bearings. k_{eq} is the bearing equivalent stiffness, d_{eq} is the bearing equivalent damping, and k_r is the effective half-rotor stiffness. The mass of this rotor is lumped to the center and represented by $2m$.

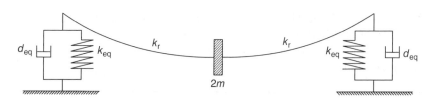

Figure 6.1   Model for the lightly and centrally loaded flexible rotor supported by two fluid-film journal bearings. From Wang and Khonsari (2006d) © ASME.

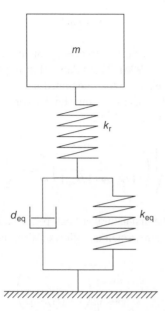

Figure 6.2   Simplified model for the rotor-bearing system. From Wang and Khonsari (2006d) © ASME.

Figure 6.2 is a simplified mass-spring-damper model for the rotor-bearing system shown in Figure 6.1. The undamped natural frequency of this simplified system is (Wang and Khonsari, 2006d)

$$\omega_{\mathrm{nd}} = \sqrt{\frac{k_{\mathrm{s}}}{m}} \qquad (6.1)$$

where k_{s} is the effective stiffness of this rotor-bearing system, $k_{\mathrm{s}} = k_{\mathrm{r}}k_{\mathrm{eq}}/(k_{\mathrm{r}} + k_{\mathrm{eq}})$.

The system stiffness has two limiting cases. If $k_{\mathrm{eq}} \gg k_{\mathrm{r}}$, the undamped natural frequency of this system is approximately equal to $\omega_{\mathrm{nd}} = \sqrt{k_{\mathrm{r}}/m}$. This implies that the rotor-bearing system could be approximated as a rigidly supported flexible rotor system. If, on the other hand, $k_{\mathrm{r}} \gg k_{\mathrm{eq}}$, the undamped natural frequency of this system is approximately equal to $\omega_{\mathrm{nd}} = \sqrt{k_{\mathrm{eq}}/m}$, meaning that the rotor-bearing system can be regarded as a rigid rotor supported by two fluid-film bearings. When k_{eq} is in a similar order of magnitude as k_{r}, the full equation of $k_{\mathrm{s}} = k_{\mathrm{r}}k_{\mathrm{eq}}/(k_{\mathrm{r}} + k_{\mathrm{eq}})$ must be applied to calculate the effective stiffness k_{s} and the undamped natural frequency ω_{nd}.

As a general application, the physical model of the rotor-bearing system shown in Figure 6.2 is used in the following analysis. The parametric equation of the rotor-bearing system is (Wang and Khonsari, 2006d)

$$k_s = m\omega_s^2 \tag{6.2}$$

where ω_s is the threshold whirling frequency of the subsynchronous vibration (Lund, 1965; Yu and Liu, 1996). This relationship implies that whirl starts with a whirling frequency ω_s equal to the undamped natural frequency of the system, ω_{nd}. Let ω_{st} represent the instability threshold rotor speed and Ω ($\Omega = \omega_s/\omega_{st}$) denote the threshold whirl ratio. Then

$$\omega_{st} = \frac{\omega_s}{\Omega} = \frac{1}{\Omega}\sqrt{\frac{k_s}{m}} = \frac{1}{\Omega}\sqrt{\left(\frac{1}{(1/k_r) + (1/k_{eq})}\right)\frac{1}{m}} = \frac{1}{\Omega}\sqrt{\left(\frac{1}{1 + (k_r/k_{eq})}\right)\frac{k_r}{m}} \tag{6.3}$$

Since $\omega_n = \sqrt{k_r/m}$ is the first natural critical speed of a simply supported flexible rotor, Equation 6.3 can be rewritten as follows (Wang and Khonsari, 2006d).

$$\omega_{st} = \frac{\omega_n}{\Omega}\sqrt{\left(\frac{1}{1 + (k_r/k_{eq})}\right)} \tag{6.4}$$

where $\omega_n = \sqrt{k_r/m}$.

The relationship between the bearing equivalent stiffness k_{eq} and the four stiffness coefficients $k_{IJ}(I,J=x,y)$ and four damping coefficients $d_{IJ}(I,J=x,y)$ is available in literature (see e.g., Lund, 1965; Yu and Liu, 1996). The stiffness coefficients and damping coefficients of fluid-film bearing can be obtained by solving the dynamic Reynolds equation under small perturbations at steady state equilibrium position with the fluid temperature equal to the effective temperature of the fluid film.

Let us define the following dimensionless parameters (Wang and Khonsari, 2006d):

$$\bar{K}_{IJ} = \frac{k_{IJ}}{W/C} \tag{6.5}$$

$$\bar{D}_{IJ} = \frac{d_{IJ}}{W/(C\omega)} \tag{6.6}$$

$$\bar{K}_{eq} = \frac{k_{eq}}{W/C} \tag{6.7}$$

$$\bar{K}_r = \frac{k_r}{W/C} \tag{6.8}$$

where C is the radial clearance and W is the load per bearing.

Equation 6.9 defines the normalized bearing equivalent stiffness $\bar{K}_{eq}$ in terms of the four normalized stiffness coefficients $\bar{K}_{IJ}(I,J=x,y)$ and four normalized damping coefficients $\bar{D}_{IJ}(I,J=x,y)$ (Lund, 1965).

$$\bar{K}_{eq} = \frac{\left(\bar{K}_{xy}\bar{D}_{yx} + \bar{K}_{yx}\bar{D}_{xy} - \bar{K}_{yy}\bar{D}_{xx} - \bar{K}_{xx}\bar{D}_{yy}\right)}{\left(\bar{D}_{xx} + \bar{D}_{yy}\right)} \tag{6.9}$$

It can be shown that the following important relationship exists between the whirl frequency ratio and the stiffness and damping coefficients (Lund, 1965; Yu and Liu, 1996).

$$\Omega = \sqrt{\frac{\left(\bar{K}_{eq} - \bar{K}_{xx}\right)\left(\bar{K}_{eq} - \bar{K}_{yy}\right) - \bar{K}_{xy}\bar{K}_{yx}}{\left(\bar{D}_{xx}\bar{D}_{yy} - \bar{D}_{xy}\bar{D}_{yx}\right)}} \tag{6.10}$$

Using Equations 6.7 and 6.8, Equation 6.4 can be rewritten as (Wang and Khonsari, 2006d)

$$\frac{\omega_{st}}{\omega_n} = \frac{1}{\Omega}\sqrt{\left(\frac{1}{1 + \left(\bar{K}_r/\bar{K}_{eq}\right)}\right)} = \frac{1}{\Omega}\sqrt{\left(\frac{\bar{K}_{eq}}{\bar{K}_{eq} + \bar{K}_r}\right)} \tag{6.11}$$

As an example to show how the fluid inlet temperature affects the system instability threshold speed, we will analyze the same rotor-bearing system whose technical specification has been defined in Table 5.4. The schematic figure of the rotor-bearing system is shown in Figure 5.8. Using Equations 6.1–6.11, the system instability threshold speed can be easily predicted in terms of Sommerfeld number (S).

Figure 6.3 illustrates the relationship between the Sommerfeld number (S) and the dimensionless threshold speed $\bar{\omega}_{st}$ ($\bar{\omega}_{st} = \omega_{st}\sqrt{C/g}$) of this rotor-bearing system. It shows that as the Sommerfeld number (S) *decreases* continuously, the threshold speed will first *decrease* slowly, and then *reach a valley* at some point with the Sommerfeld number (S) equal to about 0.8, followed by a steep *increase*. Figure 6.3 indicates that if the initial operating point of the system is located on the *left* side of the dotted line, the instability threshold speed will increase with *decreasing* Sommerfeld number (S). However, if the initial operating point of the system is located on the *right* side of the dotted line, the threshold speed will increase along with *increasing* Sommerfeld number (S).

The Sommerfeld number (S) of the rotor-bearing system can be controlled by changing the inlet fluid temperature while maintaining the other parameters unchanged. Figure 6.3 becomes more revealing if we examine the system instability directly as a function of the inlet oil temperature. This is shown in Figure 6.4.

Figure 6.4 shows that the system behavior from the viewpoint of instability is substantially different on two sides of the critical temperature corresponding to point A.

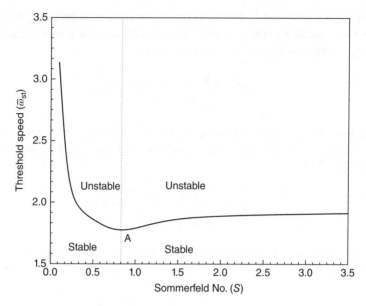

Figure 6.3   Stability of the rotor-bearing system ($L/D = 0.5$). From Wang and Khonsari (2006d) © ASME.

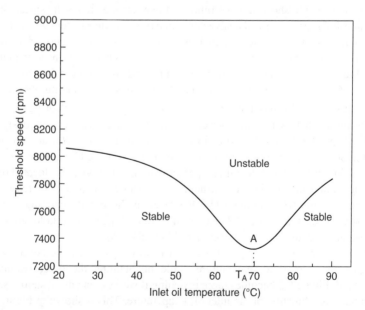

Figure 6.4   Inlet temperature effects on the instability threshold speed. From Wang and Khonsari (2006d) © ASME.

For any initial temperature T_i below $T_A \approx 69°C$ while in the range of 20–69°C, increasing the oil inlet temperature results in a drop in the instability threshold speed; for any initial temperatures T_i higher than $T_A \approx 69°C$, increasing the oil inlet temperature actually improves the instability threshold. For example, on setting the initial temperature at $T_i = 50°C$, the instability threshold speed is 7828 rpm. On increasing the inlet temperature to $T_i = 60°C$, the instability threshold speed will drop to 7556 rpm. However, if the initial temperature is $T_i = 70°C$, increasing to $T_i = 80°C$ will result in an improvement to the instability threshold speed by 248 rpm from 7329 to 7577 rpm.

The above discussion shows a theoretical basis for establishing a guideline that changing the inlet fluid temperature can control the stability of a rotor-bearing system. Next, we will present several experimental verifications for this dip phenomenon.

6.1.2 Experimental Studies

The test rig used for this study has been introduced in Section 5.2.3. It consists of a lightly and centrally loaded rotor supported by two identical hydrodynamic journal bearings on both ends. The specification of this rotor-bearing system has been given in Table 5.4. This test rig is equipped with a heating/cooling system capable of supplying oil in the range of 0–180°C. The rotor is driven by a half-horsepower AC–DC motor with an electronic speed controller through a flexible spring coupling. The motor speed range is 0–10 000 rpm. Vibration data (journal orbits) is recorded in real time by a pair of XY eddy current proximity probes installed in one of the bearing housings. These two displacement signals are processed by a computerized data acquisition system. The Bode plots, trend plots, and waterfall plots presented in this chapter are generated by this computerized data acquisition system.

Experimental tests associated with thermohydrodynamic instability require the development of a careful procedure. For this purpose, Wang and Khonsari (2006d) developed the following approach. To reach a desired inlet oil temperature, Wang and Khonsari (2006d) heated the oil and circulated it through the whole system first at a low speed—about 1800 rpm—until both the inlet oil temperature and the bearing temperature stabilized. Then, the speed was increased up to 7000 rpm at an increment of about 200 rpm. When operating above 7000 rpm, the speed increment was reduced to about 50 rpm. Especially when approaching a whirl, the speed was ramped as slowly as possible to identify the threshold speed more precisely. After crossing the threshold speed of the whirl, the speed was increased further to a higher enough value to get a sustaining whip while maintaining its amplitude at about two times the radial clearance C. Then, the speed was decreased as slowly as possible until the whirl disappeared and finally the system was allowed to coast down. In all the experiments, every time the speed was changed, the system was allowed to run long enough to reach a steady state in terms of stabilized bearing bushing temperature readings.

The existence of the hysteresis phenomenon and the important concepts of run-down threshold speed (RDTS) and run-up threshold speed (RUTS) associated with the hysteresis phenomenon introduced in Section 5.2 are particularly important to evaluate the temperature effects on system instability. In a nutshell, when a hysteresis phenomenon associated with subcritical bifurcation exists, the speed (referred to as run-down threshold speed (RDTS)) at which the whip disappears while decreasing the running speed is lower than the speed (referred to as run-up threshold speed (RUTS)) at which the whip starts.

Figure 6.5 shows the results of more than 60 sets of measurements along with the relationship between instability threshold speed and inlet oil temperature (Wang and Khonsari, 2006d). The solid line represents the theoretical predictions shown in Figure 6.4. The filled-triangle symbols represent the run-down threshold speed (RDTS), at which the whip disappears during the run-down process. The open-diamond symbols represent the run-up threshold speed (RUTS) at which the whip first starts during the run-up period. Note that the dip phenomenon is clearly evident in all experimental tests. We are now in a position to examine the dynamic behavior of this rotor-bearing system with different oil inlet temperatures.

Case No. 1: Stabilization by Reducing Oil Inlet Temperature
Let us examine a running speed ramped up to about 8050 rpm following the procedure described earlier (Wang and Khonsari, 2006d). The initial inlet oil temperature is set to 44.4°C, and the system keeps running until the bearing reaches the

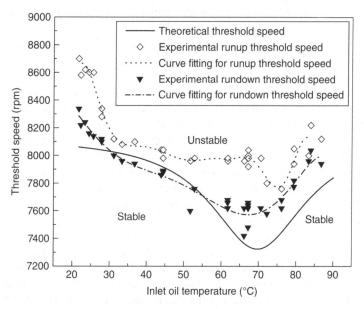

Figure 6.5 Experimental verification of the dip phenomenon. From Wang and Khonsari (2006d) © ASME.

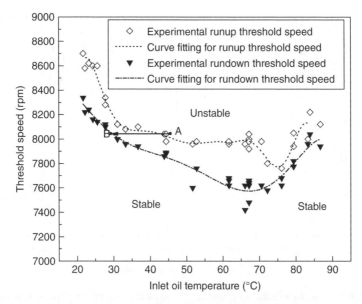

Figure 6.6 Case No. 1: Operating points change (from unstable to stable) as the inlet temperature is decreased from 44.4°C to 28.9°C. From Wang and Khonsari (2006d) © ASME.

steady state, where the bearing bushing temperature stabilizes. This operating condition referred to as point A in Figure 6.6 is unstable since the operating point A is above the curve-fitted line for RUTS. Next, the inlet temperature is decreased to 28.9°C at a decrement of 2°C, ensuring that the system reaches a steady state after each decrement of the inlet fluid temperature. Figure 6.7 shows the change in the amplitude of the journal vibration as a function of time. Since the time history corresponds to the history of the changing inlet fluid temperature, the x coordinate of Figure 6.7 can also be correlated with the inlet oil temperature that changes from 44.4°C (point A) to 28.9°C (point B). Figure 6.6 shows that the change in inlet fluid temperature from 44.4°C to 28.9°C stabilizes the rotor-bearing system since the system operating line crosses the curve-fitted line for RDTS as the inlet oil temperature is reduced. Figure 6.7 confirms the same finding. This example demonstrates that depending on the initial operating point, it is possible to stabilize the system by only decreasing the inlet fluid temperature. The waterfall plot in Figure 6.8 shows us the whole picture of the occurrence and disappearance of the subsynchronous whip. In the waterfall plot, the frequency spectra of the journal vibration at each time instant (corresponding to different oil inlet temperature in this case) are compiled together against time. The vibration frequency component at the system running frequency (rpm/60) is called synchronous vibration while the vibration frequency component at a frequency less than the system running frequency is called a subsynchronous vibration, which in this case is the oil whip. Figure 6.8 clearly shows that the whip (subsynchronous vibration) disappears with time (corresponding to the reduction of oil inlet temperature).

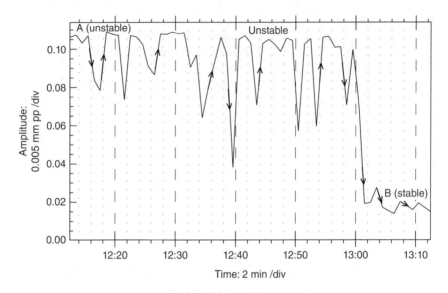

Figure 6.7   Variation of the vibration amplitude as the inlet oil temperature is decreased from 44.4°C to 28.9°C. From Wang and Khonsari (2006d) © ASME.

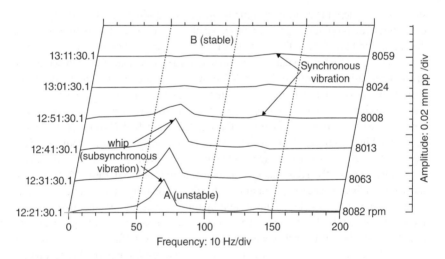

Figure 6.8   The waterfall plot of the vibration amplitude changes as inlet oil temperature is decreased from 44.4°C to 28.9°C. From Wang and Khonsari (2006d) © ASME.

Case No. 2: From Stable to Unstable System by Increasing Oil Inlet Temperature

Now let us examine the operating point at an oil inlet temperature of 31.1°C. We increase the operating speed to about 8000 rpm and let the system reach the steady state. This operating condition is stable (point A in Figure 6.9) since it is below the RDTS limit. Next, the inlet oil temperature is increased to 57.2°C at an increment of 4°C. At each increment of inlet temperature, the system is allowed to reach a steady state. This typically takes 5 minutes in this particular system.

Figure 6.10 shows the amplitude change of the journal vibration as a function of time. Again, since the history of the changing time corresponds to the history of the changing inlet temperature, the x coordinate of Figure 6.10 can also be correlated with the inlet oil temperature that changes from 31.1°C to 57.2°C. Figure 6.9 reveals the corresponding operating point change (from stable to unstable) as the inlet oil temperature is increased from 31.1°C to 57.2°C.

Figures 6.9 and 6.10 shows that, when the oil inlet temperature is at 31.1°C (point A), the rotor-bearing system is stable. When the oil inlet temperature is increased to 57.2°C (point B), the rotor-bearing system becomes unstable since the system operating line crosses the curve-fitted line for RUTS. This case shows a situation where increasing the oil inlet temperature deteriorates stability.

The above two cases focus on the left branch of the dip in the stability curve. Next we will present a set of tests that travel through the dip and into the right branch of the stability curve.

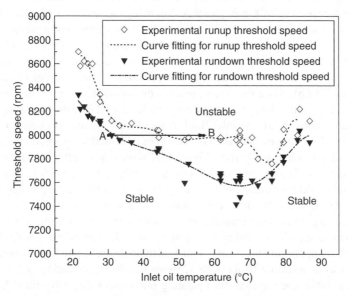

Figure 6.9 Case No. 2: Operating points change (from stable to unstable) as the inlet temperature increases from 31.1°C to 57.2°C. From Wang and Khonsari (2006d) © ASME.

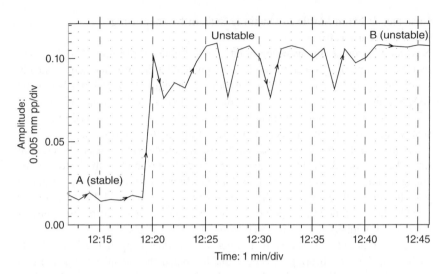

Figure 6.10   Variation of the vibration amplitude as the inlet oil temperature is increased from 31.1°C to 57.2°C. From Wang and Khonsari (2006d) © ASME.

Case No. 3: Stable System Becomes Unstable and then Resumes Stability as Oil Inlet Temperature Increases

Referring to Figure 6.11, we now begin at the oil inlet temperature of 63.9°C and increase the speed to 7800 rpm (point A) following the same procedure as the previous two cases. This initial state corresponds to the steady-state operating point A in Figure 6.11. Since the operating point A is located between RDTS and RUTS in Figure 6.11, whether the rotor-bearing system at this operating point A is stable depends on the stability condition of the previous operating point (Wang and Khonsari, 2006b). Since the system reaches this operating point A starting from an operating point below the curve-fitted line for the RDTS, this operating point A remains stable. Next, the inlet fluid temperature is increased to 88.3°C at an increment of 2°C. Again after each increment of inlet fluid temperature, the system is allowed to reach a steady state. Figure 6.12 shows the amplitude change of the journal orbit as a function of time. Examination of Figures 6.11 and 6.12 shows that at the initial inlet temperature of 63.9°C (point A), the rotor-bearing system is stable. However, when the inlet temperature is increased to about 73°C (point B) while keeping the system operating speed constant, the rotor-bearing system starts to whip as soon as the system operating line crosses the curve-fitted line for RUTS during the ramping-up process of the inlet oil temperature. Upon further increasing the oil inlet temperature to about 88°C (point C), the rotor-bearing system again becomes stable since the system operating line crosses the curve-fitted line for RDTS. Figure 6.11 provides a complete illustration of the corresponding operating point changes (from stable to unstable and then stable) while increasing the inlet oil

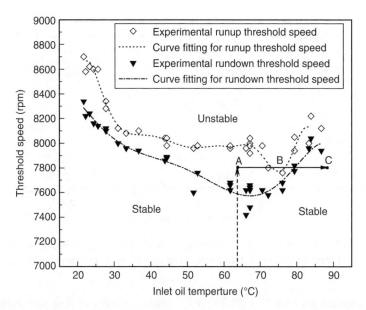

Figure 6.11 Case No. 3: Operating points change (from stable to unstable then stable) as the inlet temperature is increased from 63.9°C to 88.3°C. From Wang and Khonsari (2006d) © ASME.

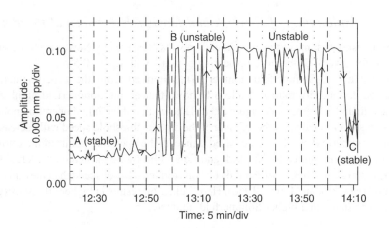

Figure 6.12 The trend plot of the vibration amplitude changes as inlet oil temperature is increased from 63.9°C to 88.3°C. From Wang and Khonsari (2006d) © ASME.

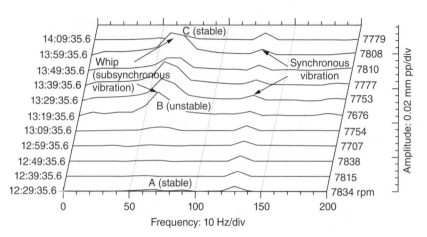

Figure 6.13 The waterfall plot of the vibration amplitude changes as inlet oil temperature is increased from 63.9°C to 88.3°C. From Wang and Khonsari (2006d) © ASME.

temperature from 63.9°C to 88.3°C. Examination of the waterfall plot shown in Figure 6.13 confirms these findings. It shows that the whip starts around 1 p.m., lasts for a period of time (about 65 minutes), and then disappears. This example shows a situation where increasing the inlet temperature up to a point deteriorates stability, but a further increase in the inlet fluid temperature resumes stability.

6.1.3   Explanation of Newkirk and Lewis's Experimental Results

Table 6.1 shows some of the experimental results presented by Newkirk and Lewis (1956) for several different rotor-bearing systems with $L/D = 0.5$.

In Table 6.1, Sommerfeld number (S) is derived from the experimentally obtained eccentricity ratio ε. Table 6.1 shows that all the Sommerfeld numbers are less than 0.3. Remember Figure 6.3 has shown that if the initial operating point of the system is located on the left side of the dip, the threshold speed will decrease as the Sommerfeld number is increased. Since all the operating points of the rotor-bearing systems studied by Newkirk and Lewis (1956) are located on the left side of the dip in Figure 6.3, a whirl builds up "on cold oil at lower speed than they do on hot oil" (Newkirk and Lewis, 1956). Here, colder oil results in higher oil viscosity, which translates to a higher Sommerfeld number.

6.1.4   Design Guidelines for Improving System Stability Based on Oil Supply Temperature

We have now established that the stability of a rotor-bearing system is a strong function of the oil inlet temperature and it is possible to move the operating point of a rotor-bearing system to a stable region by either heating or cooling the oil supply

Table 6.1 Experimental results given for different rotor-bearing systems (Newkirk and Lewis, 1956)

Different rotor-bearing systems	Oil viscosity $\mu \times 0.0069$ Pa·s	Threshold speed ω_{st} (rpm)	Eccentricity ratio ε	Sommerfeld No. S
Rotor No.1 supported	3.45	3375	0.57	0.295
by bearing No. "7-1-2"	2.12	3600	0.62	0.222
	1.42	4100	0.68	0.151
	0.85	5900	0.67	0.162
Rotor No. 1 supported	5.24	4600	0.58	0.279
by bearing No. "9-1-4"	3.58	5900	0.60	0.250
	2.99	7000	0.62	0.222
	2.28	7900	0.64	0.197
Rotor No. 3 supported	2.76	2600	0.80	0.055
by bearing No. "7-1-2"	1.72	3250	0.82	0.044
	1.06	3500	0.85	0.030
	0.72	4200	0.89	0.016
Rotor No. 3 supported	7.30	2250	0.82	0.044
by bearing No. "9-1-4"	2.80	2550	0.90	0.013
	1.68	2700	0.93	0.006
	0.92	3175	0.95	0.003

temperature. This, of course, depends on the initial operating point within the system stability plot. The following guidelines can be summarized from the analytical and experimental results presented in this section (Wang and Khonsari, 2006d).

1. Changing the oil supply/inlet temperature can either improve or deteriorate the stability of a rotor-bearing system depending on the operating point within the system stability plot.
2. As the oil supply/inlet temperature is increased, the instability threshold speed first decreases slowly, reaches a valley at some point, and then begins to increase thereafter. Therefore, if the initial operating point of the rotor-bearing system is located on the *right* side of the dip such as the one shown in Figure 6.4, the threshold speed will *increase* as the inlet oil temperature is *increased*, thus promoting stability. However, if the initial system parameters of the rotor-bearing system are located on the *left* side of the valley, the threshold speed will *increase* as the inlet oil temperature is *reduced*.

6.2   Effects of Inlet Pressure and Inlet Position

Fluid inlet pressure and inlet position effects on the static performance of axially grooved journal bearings were discussed in Section 1.3.3. It was shown that fluid inlet pressure and inlet position have pronounced effects on the static oil film geometry, pressure distribution, and steady-state journal position in axially grooved journal bearings (Wang and Khonsari, 2008a).

This section will investigate the effects of fluid inlet pressure and its position on the hydrodynamic instability of a rigid rotor symmetrically supported by two identical, axially - grooved journal bearings. For analytical simplicity, we assume that the bearing can be treated as infinitely long.

6.2.1 Equations of Motion with Consideration of Inlet Pressure and Position Effects

The rotor-bearing system illustrated in Figure 2.1a will be used to address the fluid inlet pressure and position effects. In Section 2.1, the equations of motion of this rotor-bearing system are given as Equations 2.3 and 2.4. These two coupled second-order equations of motion are decomposed into the four first-order equations described by Equations 2.31. For convenience, the decomposed equations of motion for long journal bearings with Reynolds–Floberg–Jakobsson (RFJ) boundary conditions are repeated below:

$$\dot{x}_1 = x_2$$

$$\dot{x}_2 = x_1 x_4^2 + \frac{1}{\bar{\omega}^2}\cos x_3$$

$$- \frac{6\Gamma_L}{\bar{\omega}}\left[\frac{(1-2x_4)x_1(\cos\alpha_c - \cos\alpha_s)^2}{(1-x_1^2)(1-x_1\cos\alpha_c)}\right.$$

$$\left. + 2x_2 \frac{(\alpha_c - \alpha_s + \sin\alpha_s\cos\alpha_s)(1-x_1\cos\alpha_c) + (x_1\cos^2\alpha_s - 2\cos\alpha_s + \cos\alpha_c)\sin\alpha_c}{(1-x_1^2)^{3/2}(1-x_1\cos\alpha_c)}\right]$$

$$\dot{x}_3 = x_4$$

$$\dot{x}_4 = - \frac{2x_2 x_4}{x_1} - \frac{1}{\bar{\omega}^2 x_1}\sin x_3$$

$$+ 6\frac{\Gamma_L}{\bar{\omega}}\left\{\frac{(1-2x_4)}{(1-x_1^2)^{3/2}(1-x_1\cos\alpha_c)}[(1+2x_1\cos\alpha_c)(\alpha_c - \alpha_s)\right.$$

$$-2(x_1 + \cos\alpha_c)(\sin\alpha_c - \sin\alpha_s) + (\sin\alpha_c\cos\alpha_c - \sin\alpha_s\cos\alpha_s)]$$

$$+ \frac{2x_2}{x_1(1-x_1^2)^2(1-x_1\cos\alpha_c)}[2x_1(\cos\alpha_c - x_1) - (2x_1\cos\alpha_s + \sin^2\alpha_s)(1-x_1\cos\alpha_c)$$

$$-\sin^2\alpha_c + 3x_1\sin\alpha_c(\alpha_c - \alpha_s) + (2x_1^2 + 2 - x_1\cos\alpha_s)\sin\alpha_s\sin\alpha_c]\bigg\}$$

$$(6.12)$$

where $x_1 = \varepsilon$, $x_2 = \dot{\varepsilon}$, $x_3 = \phi$, $x_4 = \dot{\phi}$, $\Gamma_L = \mu L R^3/(2mC^{2.5}g^{0.5})$, and $\bar{\omega} = \omega\sqrt{C/g}$. g is the gravitational constant and Γ_L represents the characteristic constant for a rotor-bearing system with a specified oil viscosity μ. The parameter S represents the Sommerfeld number and $\Gamma_L = (\pi/(2\bar{\omega}))S$. The relations between α_s and θ_s, α_c and θ_c, and α_i and θ_i are defined by Equations A.5–A.7. θ_s represents the oil pressure starting position and θ_c is the circumferential location where the oil film ruptures, i.e., cavitation begins. θ_i is the oil inlet position.

In Equation 6.12, α_c is determined by Equation 6.13 and α_s is given by Equation 6.14 if cavitation exists. If cavitation does not exist, Equation 6.16 is used to determine α_c within the range of $(\pi < \alpha_c \leq 2\pi + \alpha_i)$ and α_s equals $(\alpha_c - 2\pi)$. α_i is given by Equation 6.15.

$$
\begin{aligned}
3(1-2x_4)x_1\left(1-x_1^2\right)^{1/2} & \left[(2+x_1\cos\alpha_c)(\sin\alpha_c-\sin\alpha_i)-x_1\sin\alpha_i(\cos\alpha_c-\cos\alpha_i)\right. \\
& +(2\cos\alpha_c+x_1)(\alpha_i-\alpha_c)\Big]+6x_2\left[(2\cos\alpha_i-2\cos\alpha_c+x_1\sin^2\alpha_i)(1-x_1\cos\alpha_c)\right. \\
& -x_1\sin\alpha_i\sin\alpha_c(4-x_1\cos\alpha_i)-(2+x_1^2)\sin\alpha_c(\alpha_c-\alpha_i)+3x_1\sin^2\alpha_c\Big] \\
& +\bar{p}_i\left(1-x_1^2\right)^2(1-x_1\cos\alpha_c)=0
\end{aligned}
$$

$$(6.13)$$

$$
\begin{aligned}
3(1-2x_4)x_1\left(1-x_1^2\right)^{1/2} & [2(\sin\alpha_s-\sin\alpha_i)(1+x_1\cos\alpha_c)+(2\cos\alpha_c+x_1)(\alpha_i-\alpha_s) \\
& -x_1(\sin\alpha_s\cos\alpha_s-\sin\alpha_i\cos\alpha_i)]+6x_2[4x_1\sin\alpha_c(\sin\alpha_s-\sin\alpha_i) \\
& +(2-x_1\cos\alpha_s-x_1\cos\alpha_i)(\cos\alpha_i-\cos\alpha_s)(1-x_1\cos\alpha_c)-(2+x_1^2)\sin\alpha_c(\alpha_s-\alpha_i) \\
& -x_1^2\sin\alpha_c(\sin\alpha_s\cos\alpha_s-\sin\alpha_i\cos\alpha_i)]+\bar{p}_i\left(1-x_1^2\right)^2(1-x_1\cos\alpha_c)=0
\end{aligned}
$$

$$(6.14)$$

$$
\cos\alpha_i = \frac{x_1+\cos(\Theta_i-x_3)}{1+x_1\cos(\Theta_i-x_3)} \tag{6.15}
$$

$$
(1-2x_4)x_1\left(1-x_1^2\right)^{1/2}(2\cos\alpha_c+x_1)+2x_2\left(2+x_1^2\right)\sin\alpha_c = 0 \tag{6.16}
$$

The decomposed equations of motion (6.12) are of the form

$$
\dot{\mathbf{x}} = \mathbf{f}(\mathbf{x}, \bar{\omega}) \tag{6.17}
$$

The steady-state equilibrium position $\mathbf{x}_s$ in terms of $(x_{1s} = \varepsilon_s,\ x_{2s} = \dot{\varepsilon}_s,\ x_{3s} = \phi_s,$ $x_{4s} = \dot{\phi}_s)$ can be found analytically as $\mathbf{x}_s = (\varepsilon_s, 0, \phi_s, 0)$ from letting $\mathbf{f}(\mathbf{x}_s, \bar{\omega}) = 0$. By rearranging equations $\mathbf{f}(\mathbf{x}_s, \bar{\omega})|_{\mathbf{x}_s = (\varepsilon_s, 0, \phi_s, 0)} = 0$, Equations 6.18 and 6.19 can be obtained to predict ε_s and ϕ_s (Wang and Khonsari, 2008b).

$$
S = \frac{2\bar{\omega}\Gamma_L}{\pi} = \frac{\left(1-\varepsilon_s^2\right)^{1.5}(1-\varepsilon_s\cos\alpha_c)}{3\pi\varepsilon_s\sqrt{\left(1-\varepsilon_s^2\right)(\cos\alpha_c-\cos\alpha_s)^4+A_1^2}} \tag{6.18}
$$

$$
\phi_s = \tan^{-1}\left[A_1\left(1-\varepsilon_s^2\right)^{-0.5}(\cos\alpha_c-\cos\alpha_s)^{-2}\right] \tag{6.19}
$$

where $A_1 = (1 + 2\varepsilon_s\cos\alpha_c)(\alpha_c - \alpha_s) - 2(\varepsilon_s + \cos\alpha_c)(\sin\alpha_c - \sin\alpha_s) + \sin\alpha_c\cos\alpha_c - \sin\alpha_s\cos\alpha_s$.

Under different oil inlet conditions (Θ_i and $\bar{p}_i$) and with the assumption of RFJ boundary conditions, the equation of motion (Eq. 6.17) can be solved directly using the Runge–Kutta–Fehlberg method at a specified operating speed $\bar{\omega}$ with the initial velocity of the journal center set to zero. The time step size is changed adaptively according to the given error tolerance of both the relative and absolute error tolerance set to 10^{-8}. To predict the instability threshold speed $\bar{\omega}_{st}$ at a given Sommerfeld number (e.g., $S = 0.04$), the trial-and-error method presented in Section 3.2 is used to determine the initial position of the journal center, the next iterative speed $\bar{\omega}$, and then $\bar{\omega}_{st}$ based on the criteria presented in Section 3.2.1.

6.2.2 Influence of Oil Inlet Pressure on the Instability Threshold Speed

Assume oil inlet position $\Theta_i = 0$, for different dimensionless oil inlet pressure $\bar{p}_i = 0$, $\bar{p}_i = 0.5$, and $\bar{p}_i = 1.0$, the dimensionless instability threshold speeds corresponding to a series of Sommerfeld numbers are obtained using the trial-and-error method presented in Section 3.2. Figures 6.14 and 6.15 show the comparison of instability threshold speeds corresponding to different oil inlet pressures $\bar{p}_i = 0$, $\bar{p}_i = 0.5$, and $\bar{p}_i = 1.0$ as a function of Sommerfeld number and eccentricity ratio, respectively.

Figure 6.14 shows the influence of the oil inlet pressure on the instability threshold speed as a function of Sommerfeld number. The instability threshold speed

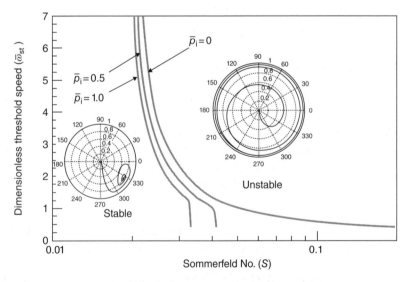

Figure 6.14 Instability threshold speeds corresponding to a series of Sommerfeld No. S with $\Theta_i = 0$. From Wang and Khonsari (2008b) © Elsevier Limited.

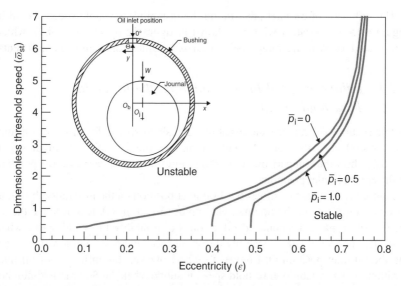

Figure 6.15 Instability threshold speeds corresponding to a series of steady state eccentricity ratio ε with $\Theta_i = 0$. From Wang and Khonsari (2008b) © Elsevier Limited.

with the oil inlet position $\Theta_i = 0$ *decreases* with *increasing* the oil inlet pressure from 0 to 1.0.

Figure 6.15 shows the influence of oil inlet pressure on the instability threshold speed as a function of steady-state eccentricity ratio ε with oil inlet position at $\Theta_i = 0$. The instability threshold speed with oil inlet position $\Theta_i = 0$ *decreases* with *increasing* the oil inlet pressure from 0 to 1.0. For oil inlet pressure $\bar{p}_i = 0.5$, the instability threshold speed drops rapidly to zero when the steady-state eccentricity ratio ε decreases down to about 0.4. For oil inlet pressure $\bar{p}_i = 1.0$, the instability threshold speed drops rapidly to zero when the steady-state eccentricity ratio ε decreases down to about 0.49. These results are consistent with the static performance shown in Figures 1.7, 1.8, 1.9, and 1.10. As Figures 1.7, 1.8, 1.9, and 1.10 show that under the oil inlet condition ($\Theta_i = 0$, $\bar{p}_i = 0.5$), a full 2π oil film exists in the fluid-film journal bearing and the steady-state attitude angle $\phi = 90°$ when $\varepsilon < 0.4$. With oil inlet condition of ($\Theta_i = 0$, $\bar{p}_i = 1.0$), a full 2π oil film exists in the fluid-film journal bearing and the steady-state attitude angle $\phi = 90°$ if the steady-state eccentricity ratio $\varepsilon < 0.49$. A rotor-bearing system supported by the full 2π oil film journal bearings with the steady-state attitude angle $\phi = 90°$ is inherently unstable. The results shown in Figures 6.14 and 6.15 also show that the instability threshold speed of a lightly loaded rotor-bearing system is sensitive to oil inlet pressure. Therefore, the operations of these pressurized bearings should be limited to a relatively heavy-loaded application.

The influence of oil inlet pressure on the instability threshold speed shown in Figure 6.15 is consistent with that shown in figure 12 by Zhang (1989), which is predicted based on the traditionally linearized stiffness and damping coefficients.

6.2.3 Influence of Oil Inlet Position on the Instability Threshold Speed

In this section we examine the instability threshold speeds for different oil inlet positions ($\Theta_i = 0°$, 45°, and 90°) with the oil inlet pressure of $\bar{p}_i = 0$. Figures 6.16 and 6.17 show the comparison of instability threshold speed curves corresponding to different oil inlet positions.

Figure 6.16 shows the influence of oil inlet position on the instability threshold speed as a function of the Sommerfeld number S with $\bar{p}_i = 0$. It shows that the oil inlet position has a pronounced effect on the instability threshold speed when $S > 0.05$. For $S > 0.05$, the instability threshold speed increases along with increasing the oil inlet position Θ_i from 0° to 90°. However, the influence of oil inlet position on the instability threshold speed is subtle when the Sommerfeld number $S < 0.03$.

Figure 6.17 shows the influence of oil inlet position on the instability threshold speed as a function of steady-state eccentricity ratio ε with $\bar{p}_i = 0$. It indicates that with oil inlet pressure $\bar{p}_i = 0$, when the steady-state eccentricity ratio ε is less than around 0.35, the instability threshold speed *increases* with *increasing* the oil inlet position Θ_i from 0° to 90°; when the steady-state eccentricity ratio ε is greater than around 0.64, the instability threshold speed *decreases* with *increasing* the oil

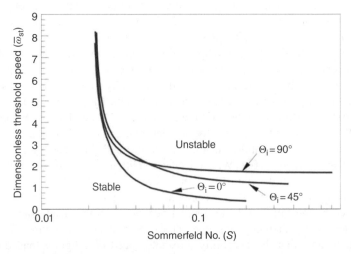

Figure 6.16 Instability threshold speeds corresponding to a series of Sommerfeld No. S with $\bar{p}_i = 0$. From Wang and Khonsari (2008b) © Elsevier Limited.

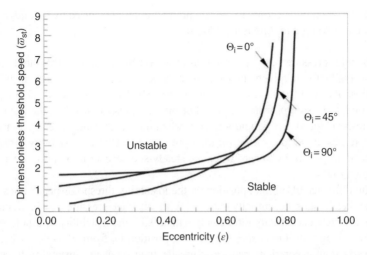

Figure 6.17 Instability threshold speeds corresponding to a series of steady-state eccentricity ratio ε with $\bar{p}_i = 0$. From Wang and Khonsari (2008b) © Elsevier Limited.

inlet position Θ_i from $0°$ to $90°$; when the steady-state eccentricity ratio ε is between 0.35 and 0.64, the effect of oil inlet position on the instability threshold speed becomes subtle and complicated.

The oil inlet position effects shown in Figure 6.17 are in agreement with those shown in figure 85 by Lundholm (1971), which are predicted based on the traditionally linearized stiffness and damping coefficients. The figure 85 (Lundholm, 1971) shows that when the steady-state eccentricity ratio ε is less than 0.3, the instability threshold is increased along with *increasing* the oil inlet position Θ_i from $0°$ to $90°$; when the steady-state eccentricity ratio ε is greater than 0.65, *increasing* the oil inlet position Θ_i from $0°$ to $90°$ lowers the instability threshold; when the steady-state eccentricity ratio ε is between 0.3 and 0.65, the effect of oil inlet position on the instability threshold becomes subtle and complicated.

Figures 6.14, 6.15, 6.16, and 6.17 show that, with oil inlet position $\Theta_i = 0$ and light load (light load results in high Sommerfeld No. and low eccentricity ratio ε), the instability threshold speed of long journal bearing approaches zero. Thus, generally speaking, for applications with oil inlet position $\Theta_i = 0$ and light load, a short bearing length is recommended. For applications with heavy load, a longer bearing length is recommended since it has a higher load-carrying capacity and a medium to high eccentricity can be obtained.

6.2.4 Design Guidelines on Inlet Pressure and Inlet Position

When specifying the oil inlet pressure and inlet position for a rotor-bearing system especially when long journal bearings are involved, the following guidelines should be taken into account to prevent premature failure caused by hydrodynamic instability (Wang and Khonsari, 2006b).

1. Oil inlet pressure has a pronounced effect on the instability threshold speed, especially for relatively low steady-state eccentricity ratio ε. The instability threshold speed with oil inlet position $\Theta_i = 0$ *decreases* with *increasing* the oil inlet pressure $\bar{p}_i$ from 0 to 1.0. For oil inlet pressure $\bar{p}_i = 0.5$, the instability threshold speed drops rapidly to zero when the steady-state eccentricity ratio ε decreases down to 0.4. For oil inlet pressure $\bar{p}_i = 1.0$, the instability threshold speed drops rapidly to zero when the steady-state eccentricity ratio ε decreases down to 0.49.
2. The influence of oil inlet position on the instability threshold speed varies with the steady-state eccentricity ratio ε. With oil inlet pressure $\bar{p}_i = 0$, when the steady-state eccentricity ratio ε is less than 0.35, the instability threshold speed *rises* along with *increasing* the oil inlet position Θ_i from $0°$ to $90°$; when the steady-state eccentricity ratio ε is greater than 0.64, the instability threshold speed *decreases* with *increasing* the oil inlet position Θ_i from $0°$ to $90°$; when the steady-state eccentricity ratio ε is between 0.35 and 0.64, the effect of oil inlet position on the instability threshold speed is subtle and complicated.

It follows, therefore, that in applications where long journal bearings with oil inlet position $\Theta_i = 0$ are specified, the lower the oil inlet pressure, the higher the instability threshold speed becomes.

Since the influence of oil inlet position on the instability threshold speed varies with the steady-state eccentricity ratio ε, to specify an optimized oil inlet position Θ_i and oil inlet pressure $\bar{p}_i$ for a long journal bearing without knowing the approximate range where the steady-state eccentricity ratio ε is, the corresponding instability threshold speed should be checked by solving the analytical equation of motion (i.e., Eq. 6.17) using the trial-and-error method. If it is known that the long journal bearing will be loaded with a light load, moving the oil inlet position from $0°$ to $90°$ can improve the system instability threshold speed significantly.

6.3   Rotor Stiffness Effects

In the previous sections of this book, all the bifurcation analyses we have presented are based on the prediction of the equations of motion (i.e., Eq. 2.32) applicable to rigid rotors. Except in the linearized analysis for instability threshold speed presented in Section 6.1, the effect of rotor stiffness on the nonlinear hydrodynamic instability of rotor-bearing system has not been discussed. Referring to Section 5.2, where we discussed the existence and characteristics of the hysteresis phenomenon based on Hopf bifurcation analyses, we showed that neglecting rotor flexibility in the equations of motion (Eq. 2.32) results in an overestimated run-up threshold speed and less accurate hysteresis loop prediction (Wang and Khonsari, 2006b, 2006e).

Thus, to accurately predict the existence and characteristics of hysteresis phenomenon and the type of bifurcation, the rotor stiffness needs to be included in the Hopf bifurcation analyses. For this purpose, the bifurcation type of a flexible rotor supported by two fluid-film journal bearings corresponding to different dimensionless rotor stiffness $\bar{K}$ must be analyzed by applying the HBT presented in Chapter 4.

6.3.1   Equations of Motion of a Flexible Rotor

Figure 6.18 shows a centrally loaded rotor-bearing system with exaggerated deflections and system coordinates. In Figure 6.18, while the polar coordinates (ε, ϕ) are used to describe the dynamic position of the journal center O_{jd}, the Cartesian coordinates (x, y) are used to denote the dynamic position of the central disk center O_m; $\mathbf{f}_\varepsilon$ and $\mathbf{f}_\phi$ are the radial and tangential components of the fluid force applied on the shaft journal; and $\mathbf{f}_r$ represents the force couple applied on the journal by the shaft and applied on the central disk by the shaft.

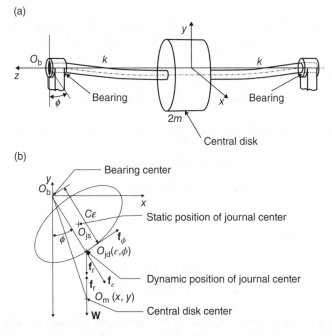

Figure 6.18   (a) Rotor-bearing system with exaggerated deflections and (b) system coordinates viewed from left side. From Wang and Khonsari (2006e) © ASME.

The following assumptions are made (Wang and Khonsari, 2006e):

1. shaft deflection is sufficiently small to permit the use of classical linear beam theory;
2. rotor mass is lumped at the midpoint (with massless shaft, central load of $2mg$, effective rotor stiffness of $2k$);
3. axial and torsional vibrations of the lumped mass are negligible;
4. gyroscopic motions of the journal and the central disk are negligible.

Let $2k$ represent the effective rotor stiffness and δ_{mid} be the static central deflection of the flexible rotor. Therefore, the effective rotor stiffness can be evaluated using (Wang and Khonsari, 2006e)

$$2k = \frac{(2mg)}{\delta_{\mathrm{mid}}} \qquad (6.20)$$

Alternatively, if the undamped natural frequency ω_n of a simply-supported and centrally-loaded rotor is known, then $2k$ can be approximated as (Wang and Khonsari, 2006e)

$$2k = 2m\omega_n^2 \qquad (6.21)$$

Based on these four assumptions and according to Newton's second law, the equations of motion for the massless journal center are (Wang and Khonsari, 2006e)

$$f_\varepsilon \sin\phi + f_\phi \cos\phi + k(x - C\varepsilon \sin\phi) = 0 \qquad (6.22)$$

$$f_\phi \sin\phi - f_\varepsilon \cos\phi + k(y + C\varepsilon \cos\phi) = 0 \qquad (6.23)$$

Solving Equations 6.22 and 6.23 for f_ε and f_ϕ yields (Wang and Khonsari, 2006e)

$$f_\varepsilon = k(-x\sin\phi + y\cos\phi + C\varepsilon) \qquad (6.24)$$

$$f_\phi = -k(x\cos\phi + y\sin\phi) \qquad (6.25)$$

According to Newton's second law, the equations of motion for the central disk are (Wang and Khonsari, 2006e)

$$m\ddot{x} + k(x - C\varepsilon \sin\phi) = 0 \qquad (6.26)$$

$$m\ddot{y} + k(y + C\varepsilon \cos\phi) + mg = 0 \qquad (6.27)$$

In the equations of motion (6.24–6.27), "." represents d/dt, and m and g are the mass per bearing and gravity constant, respectively. The parameters ε, ϕ, x, and y are all functions of time t. f_ε and f_ϕ are the radial and tangential components of the fluid force in the journal bearing, which can be determined by integrating the hydrodynamic pressure distribution obtained from the solutions to the Reynolds equation with appropriate boundary conditions applied.

Assuming that oil viscosity remains constant throughout the laminar oil film, integrating the hydrodynamic pressure distribution obtained from the solution to the Reynolds equation based on short bearing theory, and applying the half-Sommerfeld boundary conditions, we arrive at the following expressions for the fluid force components (see Chapter 1):

$$f_\varepsilon = -\frac{\mu R L^3}{2C^2}\left[\frac{2\varepsilon^2(\omega-2\dot\phi)}{(1-\varepsilon^2)^2}+\frac{\pi(1+2\varepsilon^2)\dot\varepsilon}{(1-\varepsilon^2)^{2.5}}\right] \tag{6.28}$$

$$f_\phi = \frac{\mu R L^3}{2C^2}\left[\frac{\pi(\omega-2\dot\phi)\varepsilon}{2(1-\varepsilon^2)^{1.5}}+\frac{4\varepsilon\dot\varepsilon}{(1-\varepsilon^2)^2}\right] \tag{6.29}$$

Normalizing using $\bar t = \omega t$, $\bar x = x/C$, and $\bar y = y/C$, the dimensionless form of the system equations of motion reads (Wang and Khonsari, 2006e):

$$\bar f_\varepsilon + \frac{\bar K}{\bar\omega^2}\bar x\sin\phi - \frac{\bar K}{\bar\omega^2}\bar y\cos\phi - \frac{\bar K}{\bar\omega^2}\varepsilon = 0 \tag{6.30}$$

$$\bar f_\phi + \frac{\bar K}{\bar\omega^2}\bar x\cos\phi + \frac{\bar K}{\bar\omega^2}\bar y\sin\phi = 0 \tag{6.31}$$

$$\ddot{\bar x} + \frac{\bar K}{\bar\omega^2}\bar x - \frac{\bar K}{\bar\omega^2}\varepsilon\sin\phi = 0 \tag{6.32}$$

$$\ddot{\bar y} + \frac{\bar K}{\bar\omega^2}\bar y + \frac{\bar K}{\bar\omega^2}\varepsilon\cos\phi + \frac{1}{\bar\omega^2} = 0 \tag{6.33}$$

where $\bar f_\varepsilon = -\frac{\Gamma}{\bar\omega}\left[\frac{2\varepsilon^2(1-2\dot\phi)}{(1-\varepsilon^2)^2}+\frac{\pi(1+2\varepsilon^2)\dot\varepsilon}{(1-\varepsilon^2)^{2.5}}\right]$, $\bar f_\phi = \frac{\Gamma}{\bar\omega}\left[\frac{\pi(1-2\dot\phi)\varepsilon}{2(1-\varepsilon^2)^{1.5}}+\frac{4\varepsilon\dot\varepsilon}{(1-\varepsilon^2)^2}\right]$, $\bar\omega = \omega\sqrt{C/g}$, $\bar K = k(C/mg)$, $\Gamma = \mu R L^3/(2mC^{2.5}g^{0.5})$. Both the dimensionless rotor stiffness $\bar K$ and the bearing's characteristic number Γ are constant for a specific rotor-bearing system under a specific operating condition.

The definition of the bearing's characteristic number $\Gamma = \mu R L^3/2mC^{2.5}g^{0.5}$ shows that Γ is a linear function of oil viscosity, μ. This tells us that Γ can be easily controlled by adjusting the oil viscosity μ for a given rotor-bearing system. This will be illustrated in different cases discussed in Section 6.3.3.

To solve the nonlinear equations of motion, two second-order nonlinear equations of motion (6.32–6.33) are decomposed into four first-order equations.

Let $x_1 = \varepsilon$; $x_2 = \phi$; $x_3 = \bar x$; $x_4 = \dot{\bar x}$; $x_5 = \bar y$; and $x_6 = \dot{\bar y}$. Then the equations of motion (6.30–6.33) can be rearranged as (Wang and Khonsari, 2006e)

$$\dot{x}_1 = \frac{\bar{K}\left[\pi\left(1-x_1^2\right)^{2.5}(x_3\sin x_2 - x_5\cos x_2 - x_1) + 4x_1\left(1-x_1^2\right)^2(x_3\cos x_2 + x_5\sin x_2)\right]}{\bar{\omega}\Gamma\left[\pi^2 + 2(\pi^2-8)x_1^2\right]}.$$

$$\dot{x}_2 = 0.5$$

$$+ \frac{\bar{K}\left[4\left(1-x_1^2\right)^2(x_3\sin x_2 - x_5\cos x_2 - x_1) + \pi\left(1-x_1^2\right)^{1.5}((1/x_1) + 2x_1)(x_3\cos x_2 + x_5\sin x_2)\right]}{\bar{\omega}\Gamma\left[\pi^2 + 2(\pi^2-8)x_1^2\right]}$$

$$\dot{x}_3 = x_4$$

$$\dot{x}_4 = -\frac{\bar{K}}{\bar{\omega}^2}x_3 + \frac{\bar{K}}{\bar{\omega}^2}x_1\sin x_2$$

$$\dot{x} = x_6$$

$$\dot{x}_6 = -\frac{\bar{K}}{\bar{\omega}^2}x_5 - \frac{\bar{K}}{\bar{\omega}^2}x_1\cos x_2 - \frac{1}{\bar{\omega}^2}$$

Therefore, the above system of equations is of the form:

$$\dot{\mathbf{x}} = \mathbf{f}(\mathbf{x}, \bar{\omega}) \tag{6.34}$$

The system of equations in the form of Equation 6.34 is suitable for the applications of HBT presented in Chapter 4. The steady-state equilibrium position $\mathbf{x}_s$ in terms of $(x_{1s} = \varepsilon_s, x_{2s} = \phi_s, x_{3s} = \bar{x}_s, x_{4s} = \dot{\bar{x}}_s, x_{5s} = \bar{y}_s, x_{6s} = \dot{\bar{y}}_s)$ can be defined analytically as $\mathbf{x}_s = (\varepsilon_s, \phi_s, \bar{x}_s, 0, \bar{y}_s, 0)$ by letting $\mathbf{f}(\mathbf{x}_s, \bar{\omega}) = 0$.

Letting $\mathbf{f}(\mathbf{x}_s, \bar{\omega}) = 0$, one obtains (Wang and Khonsari, 2006e)

$$\frac{x_{1s}\sqrt{16x_{1s}^2 + \pi^2\left(1-x_{1s}^2\right)}}{\left(1-x_{1s}^2\right)^2} = \frac{2}{\bar{\omega}\cdot\Gamma} \tag{6.35}$$

$$x_{2s} = \tan^{-1}\left[\frac{\pi\left(1-x_{1s}^2\right)^{0.5}}{4x_{1s}}\right] \tag{6.36}$$

$$x_{3s} = x_{1s}\sin x_{2s} \tag{6.37}$$

$$x_{4s} = 0 \tag{6.38}$$

$$x_{5s} = -x_{1s}\cos x_{2s} - \frac{1}{\bar{K}} \tag{6.39}$$

$$x_{6s} = 0 \tag{6.40}$$

where $1/\bar{K} = \delta_{\text{mid}}/C$ is the ratio of the static central deflection (δ_{mid}) of the shaft over the radial clearance (C).

Since $\bar{\omega} \cdot \Gamma = 2\pi(L/D)^2 S = S_m$, Equations 6.35–6.40 define the steady-state positions of the journal center and the central disk center if the Sommerfeld number (S) is given. Since the steady-state positions of the journal center defined by Equations 6.35 and 6.36 are consistent with those obtained by the direct solution of the Reynolds equation (Table 1.1), the position of the journal center will not change irrespective of whether the effects of the rotor stiffness are included or not. Equations 6.37–6.40 show that when the system reaches and remains in the steady-state condition, the center of the central disk is directly beneath the center of the journal with the distance equal to the static central deflection of the rotor.

6.3.2   Effects of Rotor Flexibility

As in Section 5.1 (four first-order equations of motion), the Hopf bifurcation theory presented in Chapter 4 is applied to the six first-order equations of motion (Eq. 6.34). It is used to predict the instability threshold speed and its bifurcation type corresponding to a series of bearing characteristic numbers Γ. Again, a Hopf bifurcation subroutine developed by Hassard et al. (1981) was utilized as a part of the simulation package developed by Wang and Khonsari (2006e) to obtain the six parameters ($\bar{\omega}_{st}$, γ_2, $\mathbf{v}_1$, $\beta(\bar{\omega}_{st})$, τ_2, and β_2). Appendix D shows the Jacobian matrix of the equations of motion (Eq. 6.34), which are required as one of the input variables in the developed simulation package.

Figure 6.19 shows how the dimensionless instability threshold speed and its bifurcation type change as a function of the bearing characteristic number Γ for given dimensionless values of rotor stiffness $\bar{K}$. It shows that neglecting the rotor stiffness effect can result in a misleading prediction of the instability threshold speed and its bifurcation type. For example, assume $\Gamma = 0.1$; if $\bar{K} = 0.1$, the dimensionless threshold speed $\bar{\omega}_{st} \approx 1.5$ whereas the solution for a "rigid" rotor with a huge dimensionless rotor stiffness (e.g., $\bar{K} = 10\,000$) predicts that $\bar{\omega}_{st} \approx 2.9$.

Figure 6.20 shows that for a given $\bar{K}$, how the instability threshold speed and its bifurcation type change as a function of the modified Sommerfeld number, $S_m = 2\pi(L/D)^2 S = \bar{\omega}\Gamma$. The effect of rotor stiffness $\bar{K}$ on the instability threshold speed shown in this figure is consistent with the results presented by Raimondi and Szeri (1984) based on the linear theory. In addition to the information on the instability threshold speed, Figure 6.20 can be used to determine whether a system operating at a given S_m is likely to go through subcritical bifurcation or supercritical bifurcation. However, to predict the system's instability threshold speed and its bifurcation type using Figure 6.20, one must draw the system operating line in Figure 6.20 first and then determine its crossing point with the stability curve corresponding to a given $\bar{K}$. This method will be illustrated through an example in Section 6.3.3.1.

Figure 6.21 shows how the instability threshold speed and its bifurcation type change as a function of the steady-state eccentricity ratio ε for a rotor-bearing system with a given $\bar{K}$.

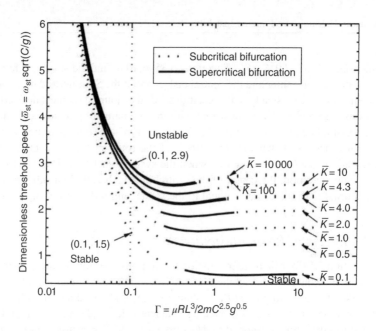

Figure 6.19 Instability threshold speed and its bifurcation type changing with increasing the bearing characteristic number Γ for a rotor-bearing system with a given dimensionless rotor stiffness $\bar{K}$. From Wang and Khonsari (2006e) © ASME.

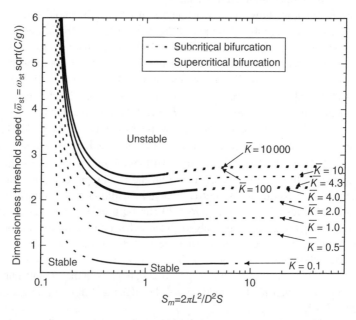

Figure 6.20 Instability threshold speed and its bifurcation type changing with increasing the modified Sommerfeld number S_m for a rotor-bearing system with a given dimensionless rotor stiffness $\bar{K}$. From Wang and Khonsari (2006e) © ASME.

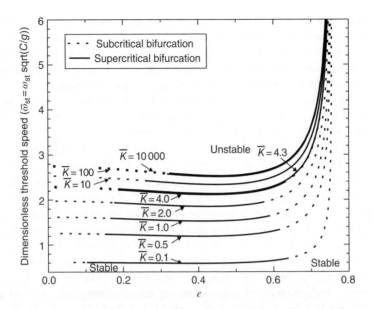

Figure 6.21   Instability threshold speed and its bifurcation type changing with increasing steady-state eccentricity ratio ε for a rotor-bearing system with a given dimensionless rotor stiffness $\bar{K}$. From Wang and Khonsari (2006e) © ASME.

Figures 6.19, 6.20, and 6.21 show that rotor stiffness has a pronounced influence on the system's instability threshold speed and its bifurcation type. Two bifurcation regions exist when the dimensionless rotor stiffness $\bar{K} \geq 4.3$. At low bearing characteristic number Γ (i.e., low modified Sommerfeld number S_m or high steady-state eccentricity ratio ε), the system goes through supercritical bifurcation. When operating at high Γ (high S_m or low ε), the bifurcation type becomes subcritical. Hollis and Taylor (1986) and Noah (1995) have reported similar results. However, three bifurcation regions are predicted to exist when the dimensionless rotor stiffness $\bar{K} < 4.3$. As the bearing's characteristic number Γ increases the bifurcation type changes from subcritical to supercritical followed by a subcritical bifurcation again. In other words, for either very high Γ (or very high S_m or very low ε) or very low Γ (or very low S_m or very high ε), the system goes through subcritical bifurcation. Therefore, supercritical bifurcation exists only when the system operates at a medium value of Γ. According to Wang and Khonsari (2006b), hysteresis phenomenon takes place only when the system goes through subcritical bifurcation. Therefore, hysteresis phenomenon can take place at either very high Γ (or very high S_m or very low ε) or very low Γ (or very low S_m or very high ε) when the dimensionless rotor stiffness is less than 4.3. These three bifurcation regions are consistent with the experimental discoveries reported by Horattas et al. (1997) involving a rotor-bearing system with bearing length over diameter ratio equal to 0.5.

When $\bar{K} < 4.3$, the three bifurcation regions make it difficult to assure that the system bifurcation is supercritical for all possible values of the bearing's characteristic number Γ. For a given rotor-bearing system, since Γ is a linear function of oil viscosity μ, any possible change in the oil viscosity affects Γ. A consequence of this kind of situation will be illustrated by an example in Section 6.3.3.

The intricacy associated with the bifurcation type puts forth a very important limitation to the well-accepted theory that higher loading is always preferred from the point view of the instability of fluid film journal bearings. For example let us assume that the dimensionless rotor stiffness $\bar{K} = 2.0$ and bearing's characteristic number $\Gamma = 0.09$ (corresponding to $\varepsilon \approx 0.7$). In this case, the dimensionless instability threshold speed is predicted to be $\bar{\omega}_{st} = 2.7$ and its bifurcation type according to Figure 6.19 or 6.21 is subcritical. According to the well-accepted linearized theory—which does not provide any information on the bifurcation type—the system will be stable even when the system's running speed is close to but still less than the instability threshold speed. But according to the Hopf bifurcation theory, this may present a dangerous situation since the system is operating in the subcritical bifurcation region. With the subcritical bifurcation, an unexpected disturbance with an amplitude outside stability envelope can easily provoke instability even when the system's running speed is less than the instability threshold speed (Wang and Khonsari, 2006c).

6.3.3 Comparison with the Results Based on Rigid-Rotor Model

The equations of motion in the case of a rigid rotor supported on two identical fluid-film journal bearings are given by Equations 2.3 and 2.4. To compare with the flexible system model given in Section 6.3.1, they are rewritten as Equations 6.41 and 6.42 for convenience:

$$\ddot{\varepsilon} - \varepsilon\dot{\phi}^2 - \bar{f}_\varepsilon - \frac{1}{\bar{\omega}^2}\cos\phi = 0 \tag{6.41}$$

$$\ddot{\phi} + \frac{2\dot{\varepsilon}\dot{\phi}}{\varepsilon} - \frac{\bar{f}_\phi}{\varepsilon} + \frac{1}{\bar{\omega}^2\varepsilon}\sin\phi = 0 \tag{6.42}$$

where all the notations are the same as those in Equations 6.30–6.33.

As discussed in Sections 2.2 and 5.1, the instability threshold speed and its bifurcation type can be determined by decomposing Equations 6.41 and 6.42 into four first-order ordinary differential equations first and then applying HBT to these resulting equations. Figure 6.22 shows that the characteristic of the system based on the flexible rotor model with a very huge dimensionless rotor stiffness $\bar{K}$ ($\bar{K} = 10\,000$) coincides with the results of a rigid rotor assumption. Generally, for $\bar{K} > 100$, the assumption of a rigid rotor should hold.

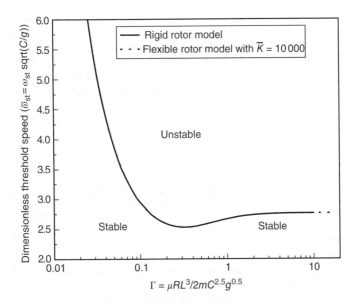

Figure 6.22   Comparison between the rigid and flexible rotor models

6.3.4   Experimental Verification

The test rig described in Section 5.2 is used for this experimental verification. It consists of a centrally loaded rotor symmetrically supported by two identical plain journal bearings on both ends. The specifications of this rotor-bearing system are given in Table 5.4. With the calculated static central deflection δ_{mid} of the flexible rotor simply supported at both ends, the dimensionless effective rotor stiffness $\bar{K}$ is estimated to be 2.9 using $k = mg/\delta_{\mathrm{mid}}$.

Based on the experimental procedure described in Section 5.2, run-up threshold speed (RUTS) and run-down threshold speed (RDTS) are identified for different oil inlet temperatures in the range of 21.2–86.7°C. If RDTS is less than RUTS, hysteresis phenomenon exists and the system goes through subcritical bifurcation. Otherwise, in the absence of hysteresis, the system goes through a supercritical bifurcation (Wang and Khonsari, 2006b).

Figure 6.23 shows a comparison between the experimental results and the stability curves presented in Figure 6.19. The *open* square denotes that the bifurcation at this operating condition is *subcritical* while *solid* square represents a *supercritical* bifurcation. The oil viscosity at the inlet temperature is used to calculate Γ. Figure 6.23 shows that the experimental results match with the stability curves presented in Figure 6.19. The stability curve corresponding to $\bar{K} = 2.9$ can be approximated through interpolation between the stability curve for $\bar{K} = 2$ and the stability curve for $\bar{K} = 4$.

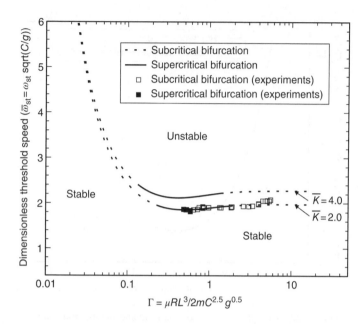

Figure 6.23   Experimental results with the given dimensionless rotor stiffness of $\bar{K} = 2.9$. From Wang and Khonsari (2006e) © ASME.

Table 6.2   Specifications of the rotor-bearing system used by Raimondi and Szeri (1984)

Journal radius, R	0.0635 m
Length of the bearing, L	0.0635 m
Radial clearance, C	127×10^{-6} m
Mass of the rotor, $2m$	2268 kg
Rotor stiffness, $2k$	8.7563×10^{8} N/m
Oil viscosity, μ	0.0138 Pa·s
Designed system's running speed, ω	5400 rpm

6.3.5   Application Examples

6.3.5.1   Illustrative Example I

Consider a rotor-bearing system described in Table 6.2 (Raimondi and Szeri, 1984). The following system constants are calculated first.

$$\bar{K} = k\left(\frac{C}{mg}\right) = 5$$

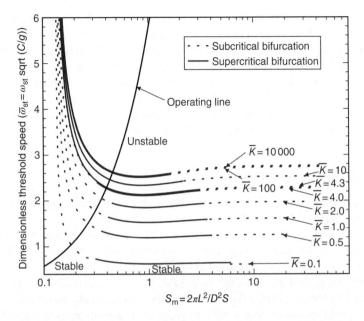

Figure 6.24   Application of Figure 6.20 on the rotor-bearing system used by Raimondi and Szeri (1984). From Wang and Khonsari (2006e) © ASME.

$$\Gamma = \frac{\mu R L^3}{\left(2mC^{2.5}g^{0.5}\right)} = 0.1737$$

Using Figure 6.19, the dimensionless instability threshold speed of this rotor-bearing system is predicted as 2.33 and the system goes through supercritical bifurcation. Using $\omega = \bar{\omega}\sqrt{g/C}$, the dimensional instability threshold speed is converted back as 6182 rpm. Thus, the system with the specified running speed of 5400 rpm is stable. These results (Figure 6.24) match very well with those predicted using figure 35 by Raimondi and Szeri (1984).

Using Equations 6.35–6.40, the steady-state equilibrium positions of the journal center and the central disk center with the system's running speed equal to 5400 rpm are predicted. When the system's running speed is at 5400 rpm, the steady-state equilibrium position of the journal center is at $(\varepsilon,\phi) = (0.6182, 45°)$ or $(\bar{x},\bar{y}) = (0.4368, -0.4374)$ and the steady-state equilibrium position of the central disk center is at $(\bar{x},\bar{y}) = (0.4368, -0.6374)$.

However, additional effort will be necessary if Figure 6.20 is selected to predict the instability threshold speed and its bifurcation type. The system's operating line must be drawn in Figure 6.20. This operating line is defined by the expression $\omega = (2mgC^2/(\mu R L^3))S_m$, which can be easily derived from the definition of modified Sommerfeld number S_m. Figure 6.24 includes the system's operating line for this example. The stability curve corresponding to $\bar{K} = 5$ can be predicted through interpolation between the stability curves for $\bar{K} = 4.3$ and $\bar{K} = 10$.

This example shows that instead of using Figure 6.20 that requires calculating the modified Sommerfeld number S_m, it is much more convenient to use Figure 6.19, which is based on the bearing's characteristic number Γ. In other words, the bearing's characteristic number Γ is a more powerful and preferred parameter for predicting the instability threshold speed and the bifurcation type of a rotor-bearing system (Wang and Khonsari, 2006e).

6.3.5.2 Illustrative Example II

An incomplete specifications of a centrally loaded rotor-bearing system used in a classical paper of Pinkus (1956) are shown in Table 6.3.

Since the locations (relative to the rotor) of the supporting bearings are not given by Pinkus (1956), suitable rotor stiffness ($\bar{K}$) is assumed so that Figure 6.19 can be used to analyze Pinkus' experimental results in 1956.

For the specified oil viscosities (0.065 Pa·s (at 25°C) and 0.014 Pa·s (at 60°C)) reported by Pinkus (1956), Table 6.4 shows the experimentally measured instability threshold speeds compared with the predicted instability threshold speeds that are based on several possible values of the dimensionless rotor stiffness $\bar{K}$. With oil viscosity equal to 0.065 Pa·s, Pinkus (1956) observed that the system could start to whip *sometimes* even at the dimensionless running speed of 1.4203.

Table 6.3 Specifications of the rotor-bearing system used by Pinkus (1956)

Journal radius, R	0.0254 m
Length of the bearing, L	0.0254 m
Radial clearance, C	63.5×10^{-6} m
Mass of the rotor, $2m$	84.823 kg

Table 6.4 Comparison of the predicted instability threshold speeds with the experimentally measured ones (Pinkus, 1956)

Oil viscosity μ (oil temperature)	Γ	Measured $\bar{\omega}_{st}$	Predicted $\bar{\omega}_{st}$ with $\bar{K} = 2$	Predicted $\bar{\omega}_{st}$ with $\bar{K} = 3$
0.065 Pa·s (25°C)	3.1699	2.0758 (1.4203)	1.9628 (subcritical)	2.1471 (subcritical)
0.014 Pa·s (60°C)	0.6827	1.8573	1.8599 (supercritical)	2.0337 (supercritical)

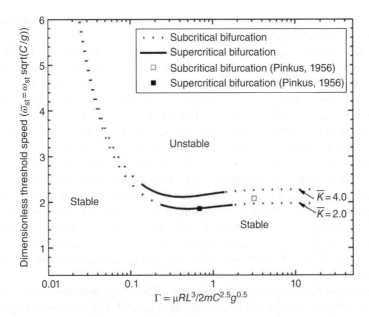

Figure 6.25   Two experimental operating points described by Table 6.4 in Figure 6.19 (Wang and Khonsari, 2006e)

Figure 6.25 marks the two experimental operating points described by Table 6.4 in Figure 6.19. In Figure 6.25, the *open* square denotes that the system goes through *subcritical* bifurcation while the *solid* square represents *supercritical* bifurcation.

Table 6.4 and Figure 6.25 show that the experimental instability threshold speeds corresponding to two different oil temperatures match with those predicted based on the assumption that the value of the dimensionless rotor stiffness $\bar{K}$ is between 2 and 3.

Table 6.4 and Figure 6.25 also show that the bifurcation that a system with oil viscosity equal to 0.065 Pa·s goes through is subcritical. Based on the Hopf bifurcation theory, even when the system's running speed is far below the instability threshold speed, a stability envelope (i.e., unstable periodic solution for the subcritical bifurcation) could exist (Wang and Khonsari, 2006c). Therefore whether the system is stable or not depends on the external perturbation that could be applied to the system. The system with a given running speed remains stable as long as the external perturbations imposed on the system are all located inside the stability envelope. Otherwise, an oil whip will start even if the external perturbation is just a transient one (Wang and Khonsari, 2006c). This kind of important information could be obtained only by using Figure 6.19, which is predicted by the Hopf bifurcation theory presented in Chapter 4. Traditional solutions based on the linear theory can not provide any information about the bifurcation type and the shape and size of the stability envelope.

Based on the above discussions, the 'peculiar' phenomenon observed by Pinkus (1956) can be explained now. The Hopf bifurcation analysis shows that, with oil viscosity equal to 0.065 Pa·s, the system goes through subcritical bifurcation. Therefore, the system is very sensitive to perturbation (Wang and Khonsari, 2006c). A transient perturbation, which often exists in practice, can easily force a stable system to become unstable even when the system's running speed is far below the predicted instability threshold speed. Therefore, when a disturbance with a certain amplitude exists, Pinkus' system when the oil viscosity is equal to 0.065 Pa·s could start to whip even at dimensionless running speeds as low as 1.4203.

Table 6.4 and Figure 6.25 also reveal that the bifurcation that the system with oil viscosity equal to 0.014 Pa·s goes through is supercritical. Since the type of the Hopf bifurcation is supercritical, stability envelope does not exist (Wang and Khonsari, 2006c). Hence, the system is always stable as long as the system's running speed is less than the instability threshold speed regardless of the external perturbation.

This example also verifies that the type of the Hopf bifurcation that a rotor-bearing system possesses can be changed from subcritical bifurcation to supercritical bifurcation by changing the oil viscosity. The change in the oil viscosity from 0.065 Pa·s (at 25°C) to 0.014 Pa·s (at 60°C) for the rotor-bearing system reported by Pinkus (1956) resulted in a change in the Hopf bifurcation type from subcritical bifurcation to supercritical bifurcation.

6.3.5.3   Illustrative Example III

In this section we analyze a centrally loaded flexible rotor symmetrically supported by two fluid-film journal bearings studied by Hori and Kato (1990) with the specifications given in Table 6.5. These bearings are typically used in generators. With the dimensionless natural frequency $\bar{\omega}_n = 1$, linear and nonlinear analyses for different B_p values have been studied by Guo and Adams (1995), and Adams and Guo (1996). The instability threshold speeds predicted by the nonlinear analysis based on the direct solution of the equations of motion using the Runge–Kutta-Felhberg

Table 6.5 Specifications of the journal bearings used by Guo and Adams (1995) and Adams and Guo (1996)

Journal diameter, $2R$	0.662 m
Bearing length, L	0.331 m
Radial clearance, C	0.000993 m
Oil viscosity, μ	0.0056 Pa·s

Table 6.6 Simulation results given by Guo and Adams (Guo, 1995; Guo and Adams, 1995; Adams and Guo 1996) with $\bar{K} = 1$

B_p	$\omega_{st}\sqrt{C/g}$	Ratio of RUTS over ω_c	Ratio of RDTS over ω_c	Hysteresis exists, yes/no?
0.5	1.026	2.00	2.00	No
0.2	1.209	1.92	1.85	Yes
0.1	1.779	2.36	1.72	Yes
0.05	2.688	3.48	1.73	Yes

method are consistent with those predicted by the linear analysis (Guo and Adams, 1995; Adams and Guo, 1996).

Table 6.6 shows the data collected from tables 3.5.1, 3.5.2, and 4.3.1 in Guo (1995) and the other figures and data in Guo and Adams (1995) and Adams and Guo (1996).

In Table 6.6, B_p is defined as (Guo and Adams, 1995; Adams and Guo, 1996):

$$B_p = \frac{\mu L}{\pi m}\left(\frac{R}{C}\right)^3\sqrt{\frac{C}{g}} = \frac{S}{\omega\sqrt{C/g}} = \frac{S}{\bar{\omega}} \qquad (6.43)$$

From the definition of Γ, we have

$$\Gamma = 2\pi\left(\frac{L}{D}\right)^2\frac{S}{\bar{\omega}} \qquad (6.44)$$

So, the relation between Γ and B_p can be described as

$$\Gamma = 2\pi\left(\frac{L}{D}\right)^2 B_p \qquad (6.45)$$

Since $\bar{\omega}_n = 1$ for the rotor-bearing system used by Guo and Adams (1995), and Adams and Guo (1996), the dimensionless rotor stiffness is

$$\bar{K} = \frac{Ck}{mg} = \frac{C\omega_n^2}{g} = \left(\omega_n\sqrt{\frac{C}{g}}\right)^2 = \bar{\omega}_n^2 = 1$$

Based on the conclusion that the hysteresis phenomenon only exists when the system undergoes subcritical bifurcation, further information summarized in Table 6.7 can be obtained.

Figure 6.26 shows that the results reported by Guo and Adams (1995), and Adams and Guo (1996) match well with the stability curve presented in

Table 6.7   Further information about the system with $\bar{K} = 1$

B_{p}	Γ	$\omega_{\mathrm{st}}\sqrt{C/g}$	Bifurcation type
0.5	0.7854	1.026	Supercritical
0.2	0.3142	1.209	Subcritical
0.1	0.1571	1.779	Subcritical
0.05	0.0785	2.688	Subcritical

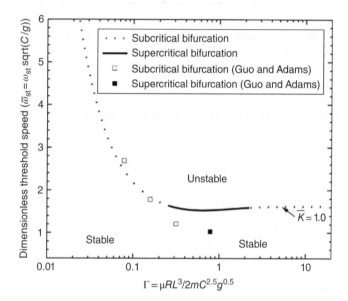

Figure 6.26   Comparison of the results reported by Guo and Adams (1995; Adams and Guo, 1996) with the stability curve presented in Figure 6.19 with $\bar{K} = 1$

Figure 6.19 when $\bar{K} = 1$. In Figure 6.26, the *open* square denotes that the bifurcation that the system undergoes is *subcritical* bifurcation; the *solid* square denotes that the system will go through *supercritical* bifurcation.

This example also verifies the predictions of the instability threshold speed and its bifurcation type using the stability curves presented in Figure 6.19.

6.3.6   Design Guidelines on Rotor Stiffness

The Hopf bifurcation theory is used to study the rotor stiffness effects on the bifurcation characteristics of a flexible rotor supported by two identical fluid-film journal bearings. Bifurcation results based on a rigid rotor supported by the same fluid-film journal bearings are used as a extreme case to verify the effectiveness of the new model for a flexible rotor-bearing system.

It is shown that the rotor stiffness has a remarkable influence on the instability threshold speed as well as its bifurcation type. For short bearings, two bifurcation regions exist when the dimensionless rotor stiffness $\bar{K} \geq 4.3$. However, three bifurcation regions will exist if the dimensionless rotor stiffness $\bar{K} < 4.3$.

Using Figure 6.19, not only the instability threshold speed but also its bifurcation type can be easily identified for any given rotor-bearing system with any given set of operating parameters.

The instability threshold speed and its bifurcation type predicted by Figure 6.19 should be beneficial at the design stage as well as for troubleshooting a rotor-bearing system that suffers from instability. If the bifurcation type corresponding to the current bearing's characteristic number Γ is in the subcritical bifurcation regime, effort should be focused on changing the bearing's characteristic number Γ by controlling the oil inlet temperature or even the oil grade to force the bifurcation into a supercritical bifurcation regime. Section 6.3.3 has illustrated how to apply these design rules through several application examples.

6.4   Worn Bearing Bushing Effects

In this section we explore the influence of roundness imperfection of the bearing bushing on rotor-bearing system instability. Depending on the loading conditions and the actual material of the bearing bushing and the rotor shaft, the bearing bushing is susceptible to wear after a period of use, particularly as a result of frequent start/stop. This kind of out-of-roundness can often have significant influence on the dynamic performance of rotor-bearing systems. In this section, we examine the influence of worn bearing bushing on the performance of rotor-bearing systems.

6.4.1   Wear Profile Model

In 1983, Dufrane et al. published the result of a series of inspection and measurement data on several worn turbine journal bearings (see the measured data in Figure 6.27). They showed that the wear pattern is concentrated nearly symmetrically at the bottom of the bearing and uniformly in the axial direction. They also proposed two different physical models to fit the measured wear profile. One of them was called footprint wear model; the other was named as abrasive wear model. By reproducing the measured experimental wear data in figure 5 presented by Dufrane et al. (1983), Figure 6.27 shows the comparison of the measured experimental data and these two mathematical models. In addition, Figure 6.27 also shows a new model proposed by Wang and Khonsari in this book to better represent the measured wear profile described by Dufrane et al. (1983).

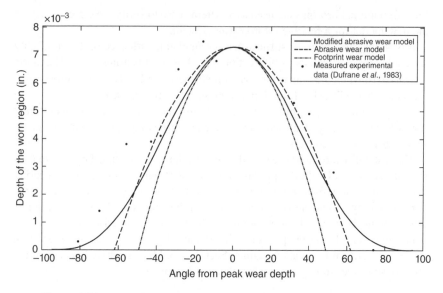

Figure 6.27   Three different models to fit the measured bushing wear profile

6.4.1.1   Model 1: Footprint Wear Model

Dufrane *et al.* (1983) postulated that the journal imprints itself into the bearing bushing leaving a footprint that can be described as follows

$$
\delta = \begin{cases} -C\left(1 + \left(1 + \dfrac{\delta_0}{C}\right)\cos(\theta + \phi)\right) & \theta_s < \theta < \theta_e \\ 0 & \text{otherwise} \end{cases}
\tag{6.46}
$$

where θ_s is the position from which wear starts, θ_e is the position at which wear ends, δ_0 is the maximum depth of the wear, C is the radial clearance, and ϕ is the attitude angle of the journal. The same symbol will be used in two additional models described next.

6.4.1.2   Model 2: Hypothetical Abrasive Wear Model

Based on a hypothetical abrasive wear model with the worn arc at a radius larger than the journal, the following model for the wear depth was also given by Dufrane *et al.* (1983):

$$
\delta = \begin{cases} \delta_0 - C(1 + \cos(\theta + \phi)) & \theta_s < \theta < \theta_e \\ 0 & \text{otherwise} \end{cases}
\tag{6.47}
$$

After comparing the measured data with the above two wear models, Dufrane *et al.* concluded that the hypothetical abrasive wear model "is probably more realistic because it allows the wear to spread beyond the footprint region." They reasoned, "This spreading would be expected because of the combination of the abrasion and slight lateral motion of the journal in the bearing." The hypothetical abrasive wear model is found to be a better representation of the measured wear profile than the footprint wear model. Based on this conclusion, the hypothetical abrasive wear model has been adopted by many different researchers in the analysis of the wear effects on the hydrodynamic lubrication (Dufrane *et al.*, 1983; Hashimoto *et al.*, 1986a, 1986b; Vaidyanathan and Keith, 1991; Kumar and Mishra, 1996a, 1996b; Tanaka and Suzuki, 2002; Fillon and Bouyer, 2004).

However, upon further review of the original figure 5 in Dufrane *et al.* (1983) and the reproduced measurement data shown in Figure 6.27, the measured wear profile shows that there are smooth transitions between the worn and unworn regions. Neither the hypothetical abrasive wear model nor the footprint wear model comes even close to the true representation of these smooth transitions. From this point of view, the following modified abrasive wear model is suggested.

6.4.1.3 Model 3: Modified Abrasive Wear Model

Now based on a hypothetical abrasive wear model with the worn arc at a radius larger than the journal and the characteristics of the two ends of the real wear region, which is that the wear region should be smoothly connected with the unworn region of the bushing, the following modified abrasive wear model for the wear depth is proposed:

$$\delta = \begin{cases} \delta_0 \cos^3\left(\dfrac{C(\pi-\theta-\phi)}{2\delta_0}\right) & \theta_s < \theta < \theta_e \\ 0 & \text{otherwise} \end{cases} \qquad (6.48)$$

Figure 6.27 compares the measured bushing wear profiles and the above three mathematical models. It shows that the new modified abrasive wear model more accurately represents the measured bushing wear profile reported by Dufrane *et al.* (1983).

Therefore, the modified abrasive wear model will be used in the following analysis of the wear effects on the performance of rotor-bearing systems.

Letting the wear depth be $\delta = \delta_0 \cos^3(C(\pi-\theta-\phi)/2\delta_0) = 0$, we have

$$\theta_s = \frac{C-\delta_0}{C}\pi - \phi \qquad (6.49)$$

$$\theta_e = \frac{C+\delta_0}{C}\pi - \phi \tag{6.50}$$

Usually, the wear range is $0 \le \theta_e - \theta_s = (2\delta_0/C)\pi \le \pi$ and the maximum wear depth is $0 \le \delta_0 \le 0.5C$ (Dufrane et al., 1983).

6.4.2   Dynamic Pressure Distribution in Worn Journal Bearing

Assuming constant oil viscosity throughout the oil film, the Reynolds equation for infinitely short bearings under laminar flow conditions ($a_z = 1$ and $b_z = 0$) can be further reduced from Equation 1.9 to 6.51.

$$\frac{\partial^2 p}{\partial z^2} = \frac{6\mu\omega}{h^3}\frac{\partial h}{\partial \theta} + \frac{12\mu}{h^3}\frac{\partial h}{\partial t} \tag{6.51}$$

For convenience, the flow boundary conditions, Equations 1.10 and 1.11, are repeated as Equations 6.52 and 6.53.

$$\left.\frac{\partial p}{\partial z}\right|_{z=0} = 0 \tag{6.52}$$

$$p|_{z=\pm\frac{L}{2}} = 0 \tag{6.53}$$

Integrating Equation 6.51 twice and substituting the above boundary conditions, one has

$$p = \frac{3\mu}{h^3}\left(\omega\frac{\partial h}{\partial \theta} + 2\frac{\partial h}{\partial t}\right)\left(z^2 - \frac{L^2}{4}\right) \tag{6.54}$$

The film thickness for the worn journal bearing is given by

$$h = C(1 + \varepsilon\cos\theta) + \delta = \begin{cases} C + C\varepsilon\cos\theta + \delta_0\cos^3\left(\dfrac{C(\pi-\theta-\phi)}{2\delta_0}\right) & \theta_s < \theta < \theta_e \\[4mm] C(1 + \varepsilon\cos\theta) & \text{otherwise} \end{cases} \tag{6.55}$$

Substituting Equations 6.55 and 1.14 into Equation 6.54 yields the following expression for the hydrodynamic fluid pressure:

$$p = \frac{3\mu}{C^2}\left(z^2 - \frac{L^2}{4}\right)\frac{2\dot{\varepsilon}\cos\theta - (\omega - 2\dot{\phi})\varepsilon\sin\theta + \frac{3\omega}{2}\sin\left(\frac{C(\pi-\theta-\phi)}{2\delta_0}\right)\cos^2\left(\frac{C(\pi-\theta-\phi)}{2\delta_0}\right)}{\left[1 + \varepsilon\cos\theta + \frac{\delta_0}{C}\cos^3\left(\frac{C(\pi-\theta-\phi)}{2\delta_0}\right)\right]^3}$$

where $\theta_s < \theta < \theta_e$

$$(6.56)$$

Elsewhere

$$p = \frac{3\mu}{C^2}\left(z^2 - \frac{L^2}{4}\right)\frac{\left[2\dot{\varepsilon}\cos\theta - (\omega - 2\dot{\phi})\varepsilon\sin\theta\right]}{(1 + \varepsilon\cos\theta)^3} \qquad (6.57)$$

6.4.3 Hydrodynamic Fluid Force in Worn Journal Bearing

Equations 1.17 and 1.18 define the hydrodynamic fluid force components in the radial and tangential directions.

Assuming that cavitation does not occur in the convergent half of the worn region (Hashimoto *et al.*, 1986a, 1986b), the following circumferential boundary conditions are recommended:

$$\begin{cases} p|_{\theta=0} = 0 \\ p|_{\theta_e \le \theta < 2\pi} = 0 \end{cases} \text{ if } \theta_e > \pi \qquad (6.58)$$

Otherwise

$$\begin{cases} p|_{\theta=0} = 0 \\ p|_{\pi \le \theta < 2\pi} = 0 \end{cases} \qquad (6.59)$$

For the case of $\theta_e > \pi$, substituting the boundary conditions (Eq. 6.58) into the expressions for the force components (Eqs. 1.17 and 1.18) yields

$$f_\varepsilon = R\int_{-L/2}^{L/2}\int_0^{\theta_e} p\cos\theta \, d\theta dz = R\int_{-L/2}^{L/2}\left[\int_0^{\theta_s} p\cos\theta d\theta + \int_{\theta_s}^{\theta_e} p\cos\theta d\theta\right]dz \qquad (6.60)$$

$$f_\phi = R\int_{-L/2}^{L/2}\int_0^{\theta_e} p\sin\theta \, d\theta dz = R\int_{-L/2}^{L/2}\left[\int_0^{\theta_s} p\sin\theta d\theta + \int_{\theta_s}^{\theta_e} p\sin\theta d\theta\right]dz \qquad (6.61)$$

where p is given by Equations 6.56 or 6.57, respectively.

Substituting Equations 6.56 or 6.57 into Equations 6.60 and 6.61, respectively, yields

$$f_\varepsilon = -\frac{\mu R L^3}{2C^2}\left[\int_0^{\theta_s}\frac{2\dot{\varepsilon}\cos\theta-(\omega-2\dot{\phi})\varepsilon\sin\theta}{(1+\varepsilon\cos\theta)^3}\cos\theta d\theta\right.$$

$$\left.+\int_{\theta_s}^{\theta_e}\frac{2\dot{\varepsilon}\cos\theta-(\omega-2\dot{\phi})\varepsilon\sin\theta+\frac{3\omega}{2}\sin(C(\pi-\theta-\phi)/2\delta_0)\cos^2(C(\pi-\theta-\phi)/2\delta_0)}{[1+\varepsilon\cos\theta+(\delta_0/C)\cos^3(C(\pi-\theta-\phi)/2\delta_0)]^3}\cos\theta d\theta\right]$$

$$(6.62)$$

$$f_\phi = -\frac{\mu R L^3}{2C^2}\left[\int_0^{\theta_s}\frac{2\dot{\varepsilon}\cos\theta-(\omega-2\dot{\phi})\varepsilon\sin\theta}{(1+\varepsilon\cos\theta)^3}\sin\theta d\theta\right.$$

$$\left.+\int_{\theta_s}^{\theta_e}\frac{2\dot{\varepsilon}\cos\theta-(\omega-2\dot{\phi})\varepsilon\sin\theta+\frac{3\omega}{2}\sin\left(\frac{C(\pi-\theta-\phi)}{2\delta_0}\right)\cos^2\left(\frac{C(\pi-\theta-\phi)}{2\delta_0}\right)}{\left[1+\varepsilon\cos\theta+\frac{\delta_0}{C}\cos^3\left(\frac{C(\pi-\theta-\phi)}{2\delta_0}\right)\right]^3}\sin\theta d\theta\right]$$

$$(6.63)$$

where θ_s and θ_e are given by Equations 6.49 and 6.50, respectively.

Similarly, for the case of $\theta_e \leq \pi$, substituting the boundary conditions (Eq. 6.59) into the expressions for the force components, we arrive at the following:

$$f_\varepsilon = R\int_{-L/2}^{L/2}\int_0^\pi p\cos\theta d\theta dz = R\int_{-L/2}^{L/2}\left[\int_0^{\theta_s}p\cos\theta d\theta+\int_{\theta_s}^{\theta_e}p\cos\theta d\theta+\int_{\theta_e}^\pi p\cos\theta d\theta\right]dz$$

$$(6.64)$$

$$f_\phi = R\int_{-L/2}^{L/2}\int_0^\pi p\sin\theta d\theta dz = R\int_{-L/2}^{L/2}\left[\int_0^{\theta_s}p\sin\theta d\theta+\int_{\theta_s}^{\theta_e}p\sin\theta d\theta+\int_{\theta_e}^\pi p\sin\theta d\theta\right]dz$$

$$(6.65)$$

Substituting Equations 6.56 or 6.57 into Equations 6.64 and 6.65 yields

$$f_\varepsilon = -\frac{\mu R L^3}{2C^2}\left[\int_0^{\theta_s}\frac{2\dot{\varepsilon}\cos\theta-(\omega-2\dot{\phi})\varepsilon\sin\theta}{(1+\varepsilon\cos\theta)^3}\cos\theta d\theta\right.$$

$$+\int_{\theta_s}^{\theta_e}\frac{2\dot{\varepsilon}\cos\theta-(\omega-2\dot{\phi})\varepsilon\sin\theta+\frac{3\omega}{2}\sin\left(\frac{C(\pi-\theta-\phi)}{2\delta_0}\right)\cos^2\left(\frac{C(\pi-\theta-\phi)}{2\delta_0}\right)}{\left[1+\varepsilon\cos\theta+\frac{\delta_0}{C}\cos^3\left(\frac{C(\pi-\theta-\phi)}{2\delta_0}\right)\right]^3}\cos\theta d\theta$$

$$\left.+\int_{\theta_e}^\pi\frac{2\dot{\varepsilon}\cos\theta-(\omega-2\dot{\phi})\varepsilon\sin\theta}{(1+\varepsilon\cos\theta)^3}\cos\theta d\theta\right]$$

$$(6.66)$$

$$f_\phi = -\frac{\mu R L^3}{2C^2}\left[\int_0^{\theta_s}\frac{2\dot{\varepsilon}\cos\theta-(\omega-2\dot{\phi})\varepsilon\sin\theta}{(1+\varepsilon\cos\theta)^3}\sin\theta d\theta\right.$$

$$+\int_{\theta_s}^{\theta_e}\frac{2\dot{\varepsilon}\cos\theta-(\omega-2\dot{\phi})\varepsilon\sin\theta+\dfrac{3\omega}{2}\sin\left(\dfrac{C(\pi-\theta-\phi)}{2\delta_0}\right)\cos^2\left(\dfrac{C(\pi-\theta-\phi)}{2\delta_0}\right)}{\left[1+\varepsilon\cos\theta+\dfrac{\delta_0}{C}\cos^3\left(\dfrac{C(\pi-\theta-\phi)}{2\delta_0}\right)\right]^3}\sin\theta d\theta$$

$$\left.+\int_{\theta_e}^{\pi}\frac{2\dot{\varepsilon}\cos\theta-(\omega-2\dot{\phi})\varepsilon\sin\theta}{(1+\varepsilon\cos\theta)^3}\sin\theta d\theta\right]$$

(6.67)

where θ_s and θ_e are given by Equations 6.49 and 6.50, respectively.

6.4.4 Example Showing the Worn Bearing Bushing Profile and Its Pressure Profile

Table 6.8 shows the specifications of a sample rotor-bearing system. At steady state ($\dot{\varepsilon}=0$, $\dot{\phi}=0$), Equations 6.55–6.57 are applied to the rotor-bearing system to evaluate the fluid-film profile and the hydrodynamic pressure profiles with and without wear.

Figures 6.28 and 6.29 show the fluid film profiles in relative and absolute circumferential coordinates (coordinate definitions are given in Figure 1.4). Figure 6.30 shows the worn bearing bushing profile in polar coordinate.

Figures 6.31 and 6.32 show the hydrodynamic pressure distribution at $z=0$ (midplane) for the worn bearing in either relative or absolute circumferential coordinates, respectively. Figure 6.33 shows the hydrodynamic pressure distribution around the shaft journal in the polar coordinates. These pressure profiles are consistent with the reported results in the literature (e.g., Hashimoto et al., 1986a; Fillon and Bouyer, 2004) on the bearing bushing wear effects.

Table 6.8 Specifications of a sample rotor-bearing system

Journal diameter, D	25.4 mm
Length of journal bearing, L	12.7 mm
Mass of the rotor, $2m$	5.4523 kg
Radial clearance, C	0.0508 mm
Maximum wear depth, δ	0.0127 mm
Fluid viscosity, Pa·s	0.03
Rotating speed, rad/s	300

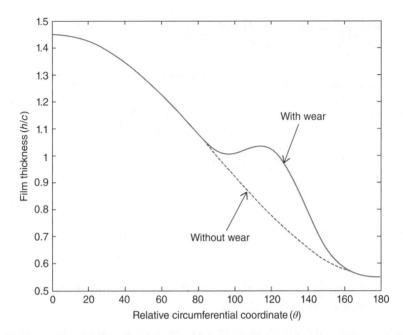

Figure 6.28   Fluid film profiles with and without wear in relative circumferential coordinates

6.4.5   Bearing Bushing Wear Effect on System Stability

Section 6.4.1 shows that the modified abrasive wear model more accurately represents the measured bearing bushing wear profile than the other two existing models. Due to the complexity of the analytical expression of the fluid force components (e.g., Eqs. 6.66 and 6.67), there is currently no closed-form solution available with the new abrasive wear model on the wear effect on the instability threshold speed of any rotor-bearing system.

On the other hand, based on the hypothetical abrasive wear model, some studies on the wear effect on the system instability threshold speed have been published using either the finite element method or the finite difference method (Hashimoto *et al.*, 1986b; Kumar and Mishra, 1996b; Tanaka and Suzuki, 2002).

Using the finite element method, Hashimoto *et al.* (1986b) discovered that, in both laminar and turbulent regimes, as soon as wear started with a plain journal bearing bushing, the instability threshold speed of the rotor-bearing system became much lower than that of the unworn plain bearing; however, as the wear depth δ_0 advanced, the system instability threshold speed improved significantly. A more interesting phenomenon was that as the wear depth δ_0 advanced further to and beyond 30% of the radial clearance C, the system instability threshold speed with heavy bearing load (low Sommerfeld number) went beyond that of the unworn

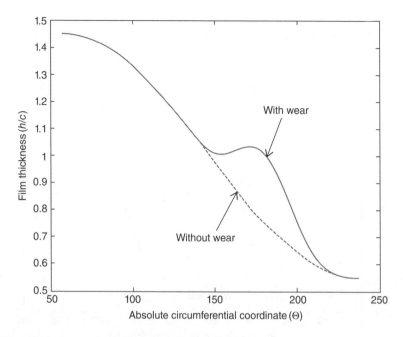

Figure 6.29   Fluid film profiles with and without wear in absolute circumferential coordinates

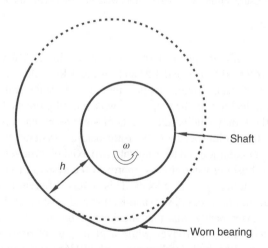

Figure 6.30   Worn bearing bushing profile

bearing while the system instability threshold speed remained lower than that of the unworn plain bearing with light load (high Sommerfeld number).

Using the finite difference method, Kumar and Mishra (1996b) confirmed these findings reported by Hashimoto *et al.* (1986b). They also reported that these same

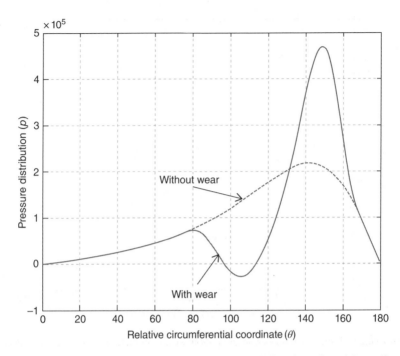

Figure 6.31   Pressure profiles with and without wear in relative circumferential coordinates

findings apply to different plain journal bearings having a different length over diameter ratio ($L/D = 0.5$, 1.0, and 1.5) and different Reynolds numbers.

In 2002, Tanaka and Suzuki applied the same hypothetical abrasive wear model to an elliptical journal bearing with two axial oil grooves. Using the finite difference method, they studied the effects of wear depth and preload factor on the instability threshold speed of the rotor-bearing system. Without preload (equivalent to a circular journal bearing with two axial oil grooves), they reported that the bearing bushing wear always improved the system instability threshold speed, especially at heavy bearing loads. This is very different from the findings reported by Hashimoto *et al.* (1986b) and Kumar and Mishra (1996b). The explanation for this discrepancy was that the journal bearing Tanaka and Suzuki adopted had two axial oil grooves while a plain journal bearing without any oil groove was used by both Hashimoto *et al.* (1986b) and Kumar and Mishra (1996b). It was also reported that as the preload factor of the elliptical bearing increased up to 0.75, the trend of the wear effect on the system instability threshold speed reversed when the preload factor crossed 0.5. When the preload factor of the elliptical bearing was equal to 0.75, the bearing bushing wear always deteriorated the system instability threshold speed. At the same time in a different

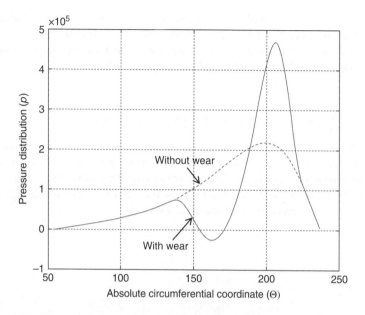

Figure 6.32   Pressure profiles with and without wear in absolute circumferential coordinates

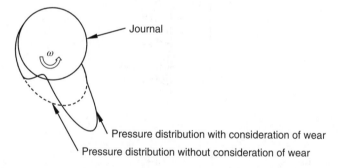

Figure 6.33   Pressure profiles with and without wear in polar coordinates

paper, Suzuki and Tanaka (2002) also published some experimental results to prove their numerical findings.

6.5   Shaft Unbalance Effects

In this section we briefly examine the effect of shaft unbalance. Certain degree of shaft unbalance always exists in a new installation as a manufacturing residual or naturally emerges during the operation as a result of uneven wear, thermal bow, sag caused by gravity and/or operational centrifugal forces, uneven multiplane centrifugal force couples, and due to the formation of uneven rust or other effects that cause surface modifications. Shaft unbalance can have significant influence on the dynamic performance of rotor-bearing systems.

6.5.1   Equation of Motion with Shaft Unbalance

Figure 6.34 shows an exaggerated deflection and two system coordinates of a centrally loaded rotor-bearing system, where shaft unbalance is included in the analysis. In Figure 6.34, O_{jd} is the dynamic position of the journal center; O_{dg} is the *geometric center* of the central disk; O_{dm} is the *mass center* of the central disk; and u is the mass eccentricity of the central disk.

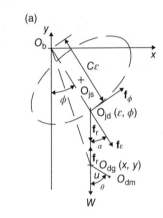

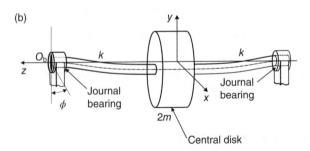

Figure 6.34   Centrally loaded rotor-bearing system with shaft unbalance. (a) System coordinates showing shaft unbalance viewed from left side. (b) Rotor-bearing system with exaggerated deflections (Wang and Khonsari, 2006e).

To analyze the dynamic performance of the rotor-bearing system with unbalance taken into account, the following six assumptions are made:

1. Shaft deflection is sufficiently small to permit the use of the classical linear beam theory;
2. The mass of the rotor is lumped at the middle of the bearing span length;
3. The mass unbalance is defined at the rotor middle point lengthwise;
4. The rotating speed is constant along the rotor;
5. The axial and torsional vibrations of the lumped mass are negligible; and
6. Gyroscopic motion of the central disk is negligible.

The equations of motion for a flexible and centrally loaded rotor horizontally supported by two identical oil-film journal bearings are given by Equations 6.22 and 6.23. These equations are repeated below for convenience:

$$f_\varepsilon \sin\phi + f_\phi \cos\phi + k(x - C\varepsilon\sin\phi) = 0$$

$$f_\phi \sin\phi - f_\varepsilon \cos\phi + k(y + C\varepsilon\cos\phi) = 0$$

Through simple algebraic rearrangement, the above equations of motion for the journal center can be rewritten as

$$f_\varepsilon = k(-x\sin\phi + y\cos\phi + C\varepsilon) \qquad (6.68)$$

$$f_\phi = -k(x\cos\phi + y\sin\phi) \qquad (6.69)$$

Based on the six assumptions made earlier, the equations of motion for the central disk are given by Equations 6.70–6.72.

$$m\ddot{x} + k(x - C\varepsilon\sin\phi) - mu\omega^2\sin\theta = 0 \qquad (6.70)$$

$$m\ddot{y} + k(y + C\varepsilon\cos\phi) + mg + mu\omega^2\cos\theta = 0 \qquad (6.71)$$

$$\dot{\theta} = \omega \qquad (6.72)$$

In the equations of motion (Eqs. 6.68–6.72), m is the rotor mass per bearing; g is gravity constant; $2k$ is the effective stiffness of the rotor. mu is the shaft unbalance per bearing in the unit of kg·m. ε, ϕ, x, and y are all functions of time t. The radial and tangential hydrodynamic fluid force components f_ε and f_ϕ were derived in Chapter 1. For short bearings with laminar flow assumptions, Equations 1.23 and 1.24 are repeated below for further derivations.

$$f_\varepsilon = -\frac{RL^3\mu}{2C^2}\left[\frac{2\varepsilon^2(\omega - 2\dot{\phi})}{(1 - \varepsilon^2)^2} + \frac{\pi(1 + 2\varepsilon^2)\dot{\varepsilon}}{(1 - \varepsilon^2)^{2.5}}\right]$$

$$f_\phi = \frac{RL^3\mu}{2C^2}\left[\frac{\pi(\omega-2\dot\phi)\varepsilon}{2(1-\varepsilon^2)^{1.5}} + \frac{4\varepsilon\dot\varepsilon}{(1-\varepsilon^2)^2}\right]$$

The equations of motion (Eqs. 6.68–6.72) are consistent with those given by Bar-Yoseph and Blech (1977).

Normalizing using $\bar{t} = \omega t$, $\bar{x} = x/C$, and $\bar{y} = y/C$, Equations 6.73–6.77 present the dimensionless forms of the equations of motion with shaft unbalance taken into account.

$$\bar{f}_\varepsilon + \lambda\bar{x}\sin\phi - \lambda\bar{y}\cos\phi - \lambda\varepsilon = 0 \tag{6.73}$$

$$\bar{f}_\phi + \lambda\bar{x}\cos\phi + \lambda\bar{y}\sin\phi = 0 \tag{6.74}$$

$$\ddot{\bar{x}} + \lambda\bar{x} - \lambda\varepsilon\sin\phi - \Delta\sin\theta = 0 \tag{6.75}$$

$$\ddot{\bar{y}} + \lambda\bar{y} + \lambda\varepsilon\cos\phi + \frac{1}{\bar{\omega}^2} + \Delta\cos\theta = 0 \tag{6.76}$$

$$\dot{\theta} = 1 \tag{6.77}$$

where $\bar{f}_\varepsilon = -\frac{\Gamma}{\bar\omega}\left[\frac{2\varepsilon^2(1-2\dot\phi)}{(1-\varepsilon^2)^2} + \frac{\pi(1+2\varepsilon^2)\dot\varepsilon}{(1-\varepsilon^2)^{2.5}}\right]$, $\bar{f}_\phi = \frac{\Gamma}{\bar\omega}\left[\frac{\pi(1-2\dot\phi)\varepsilon}{2(1-\varepsilon^2)^{1.5}} + \frac{4\varepsilon\dot\varepsilon}{(1-\varepsilon^2)^2}\right]$, $\bar\omega = \omega\sqrt{C/g}$, $\lambda = \left(\frac{\bar\omega_n}{\bar\omega}\right)^2 = \frac{k}{m\omega^2}$, $\Delta = \frac{mu}{mC}$, $\Gamma = \frac{\mu RL^3}{2mC^{2.5}g^{0.5}}$, and "." represents $d/\omega dt$.

6.5.2   Decomposition of the Equations of Motion with Shaft Unbalance

Substituting Equations 1.23 and 1.24 into the first two equations of motion (Eqs. 6.73 and 6.74) yields

$$\frac{\pi(1+2\varepsilon^2)\dot\varepsilon}{4\varepsilon^2(1-\varepsilon^2)^{0.5}} - \dot\phi = \frac{\bar\omega\lambda}{4\Gamma}\frac{(1-\varepsilon^2)^2}{\varepsilon^2}(\bar{x}\sin\phi - \bar{y}\cos\phi - \varepsilon) - 0.5 \tag{6.78}$$

$$\frac{4\dot\varepsilon}{\pi(1-\varepsilon^2)^{1/2}} - \dot\phi = -\frac{\bar\omega\lambda}{\pi\Gamma}\frac{(1-\varepsilon^2)^{1.5}}{\varepsilon}(\bar{x}\cos\phi + \bar{y}\sin\phi) - 0.5 \tag{6.79}$$

By solving $\dot\varepsilon$ and $\dot\phi$ in Equations 6.78 and 6.79, the first two equations of motion (Eqs. 6.73 and 6.74) can be rewritten as follows.

$$\dot{\varepsilon} = \frac{\bar{\omega}\lambda\left[\pi(1-\varepsilon^2)^{2.5}(\bar{x}\sin\phi-\bar{y}\cos\phi-\varepsilon)+4\varepsilon(1-\varepsilon^2)^2(\bar{x}\cos\phi+\bar{y}\sin\phi)\right]}{\Gamma[\pi^2+2(\pi^2-8)\varepsilon^2]} \tag{6.80}$$

$$\dot{\phi} = 0.5$$

$$+\frac{\bar{\omega}\lambda\left[4(1-\varepsilon^2)^2(\bar{x}\sin\phi-\bar{y}\cos\phi-\varepsilon)+\pi(1-\varepsilon^2)^{1.5}((1/\varepsilon)+2\varepsilon)(\bar{x}\cos\phi+\bar{y}\sin\phi)\right]}{\Gamma[\pi^2+2(\pi^2-8)\varepsilon^2]}$$

$$\tag{6.81}$$

To solve the nonlinear equations of motion, first, we begin by rewriting the two second-order nonlinear equations of motion (Eqs. 6.75 and 6.76) into four first-order equations.

Let $x_1 = \varepsilon$, $x_2 = \phi$, $x_3 = \bar{x}$, $x_4 = \dot{\bar{x}}$, $x_5 = \bar{y}$, $x_6 = \dot{\bar{y}}$, and $x_7 = \theta$; then the equations of motion (Eqs. 6.75–6.77, 6.80, and 6.81) can be rewritten as

$$\dot{x}_1 = \frac{\bar{\omega}\lambda\left[\pi\left(1-x_1^2\right)^{2.5}(x_3\sin x_2-x_5\cos x_2-x_1)+4x_1\left(1-x_1^2\right)^2(x_3\cos x_2+x_5\sin x_2)\right]}{\Gamma\left[\pi^2+2(\pi^2-8)x_1^2\right]}.$$

$$\dot{x}_2 = 0.5$$

$$+\frac{\bar{\omega}\lambda\left[4\left(1-x_1^2\right)^2(x_3\sin x_2-x_5\cos x_2-x_1)+\pi\left(1-x_1^2\right)^{1.5}((1/x_1)+2x_1)(x_3\cos x_2+x_5\sin x_2)\right]}{\Gamma\left[\pi^2+2(\pi^2-8)x_1^2\right]}$$

$$\dot{x}_3 = x_4$$

$$\dot{x}_4 = -\lambda x_3 + \lambda x_1 \sin x_2 + \Delta \sin x_7$$

$$\dot{x}_5 = x_6$$

$$\dot{x}_6 = -\lambda x_5 - \lambda x_1 \cos x_2 - \frac{1}{\bar{\omega}^2} - \Delta \cos x_7$$

$$\dot{x}_7 = 1$$

The above equations are of the form

$$\dot{\mathbf{x}} = \mathbf{f}(\mathbf{x}, \omega) \tag{6.82}$$

The equations of motion (Eq. 6.82) are, therefore, suitable for application of Runge–Kutta–Fehlberg method. The steady-state equilibrium position $\mathbf{x}_s$ of the rotor-bearing system in terms of $(x_{1s} = \varepsilon_s,\ x_{2s} = \phi_s,\ x_{3s} = \bar{x}_s,\ x_{4s} = \dot{\bar{x}}_s,\ x_{5s} = \bar{y}_s,\ x_{6s} = \dot{\bar{y}}_s)$ can be found analytically as $\mathbf{x}_s = (\varepsilon_s, \phi_s, \bar{x}_s, \bar{y}_s, 0, 0)$ from $\mathbf{f}(\mathbf{x}_s, \bar{\omega}) = 0$.

Letting the first six equations of $\mathbf{f}(\mathbf{x}_s, \bar{\omega}) = 0$ (except $\dot{x}_7 = 1 \neq 0$), we have

$$\frac{x_{1s}\sqrt{16x_{1s}^2 + \pi^2\left(1 - x_{1s}^2\right)}}{\left(1 - x_{1s}^2\right)^2} = \frac{2}{\Gamma\bar{\omega}} \tag{6.83}$$

$$x_{2s} = \tan^{-1}\left[\frac{\pi\left(1 - x_{1s}^2\right)^{0.5}}{4x_{1s}}\right] \tag{6.84}$$

$$x_{3s} = x_{1s}\sin x_{2s} + \left(\frac{\Delta\sin\bar{t}}{\lambda}\right) \tag{6.85}$$

$$x_{4s} = 0 \tag{6.86}$$

$$x_{5s} = -x_{1s}\cos x_{2s} - f_c - \left(\frac{\Delta\cos\bar{t}}{\lambda}\right) \tag{6.87}$$

$$x_{6s} = 0 \tag{6.88}$$

where f_c is the ratio of the static central deflection of the shaft $f_c = g/\omega_n^2 = C/\bar{\omega}_n^2$ over the radial clearance (C) and $f_c > 0$.

Since $S = \Gamma\bar{\omega}/(2\pi)(D/L)^2$, Equations 6.83–6.88 define the steady-state positions of the journal center and the central disk center when either the system characteristic number Γ and rotating speed ω or S is specified. Equations 6.83 and 6.84 are identical to those obtained by direct solution of the Reynolds equation based on the infinitely short bearing assumption presented in Chapter 1. Equations 6.85–6.88 show that when the system reaches and remains at the steady-state condition, the center of the central disk is right beneath the center of the journal.

6.5.3   Numerical Solution of the Equations of Motion

Runge–Kutta–Fehlberg method can be used to solve the system of equations (Eq. 6.82) to predict the orbits of journal and central disk. Consider a rotor-bearing system operating at a running speed $\bar{\omega}$ with a specified oil viscosity μ; the initial conditions for the nonlinear equations of motion (Eq. 6.82) are given as

$$x_1(0) = x_{10} \tag{6.89}$$

$$x_2(0) = x_{20} \tag{6.90}$$

$$x_3(0) = x_{10}\sin x_{20} \tag{6.91}$$

$$x_4(0) = 0 \tag{6.92}$$

$$x_5(0) = -x_{10}\cos x_{20} - f_c \tag{6.93}$$

$$x_6(0) = 0 \tag{6.94}$$

where x_{10} and x_{20} are the given initial eccentricity and attitude angle of the journal center.

From the discussion earlier on steady-state positions, it is reasonable to assume that the initial position of the central disk center is right beneath the initial position of the journal center with a distance equal to the static deflection at the middle of the shaft with the initial speeds $x_4(0) = x_6(0) = 0$. Similar to the stability envelope discussed in Section 5.1 for an ideal rotor-bearing system without any shaft unbalance, the selection of the initial position of the journal center can play an important role in determining the orbital stability. The time step size is changed adaptively according to the given tolerance of 10^{-10} between the fourth- and fifth-order solutions. The following example shows the shaft unbalance effects on the dynamic performance of a rotor-bearing system.

6.5.4 Example Showing Shaft Unbalance Effects on Journal Orbits

Table 6.9 shows the specifications of a sample rotor-bearing system.

By applying and solving the equation of motion (Eq. 6.82) on the rotor-bearing system with an initial 0.001 eccentricity offset release (small perturbation) from the steady-state position, the orbits of the journal and central disk can be predicted. Appendix E lists the original MATLAB code used to solve this example. Figure 6.35 shows the orbits of the journal and the central disk in the same plot. It shows that, at steady state, the central disk is located directly beneath the journal center. Figures 6.36 and 6.37 show the zoom-in view of the orbits of the journal and the central disk, respectively. These orbits show that, after being released with a tiny eccentricity offset (small perturbation) of 0.001 from the nominal steady state positions of an ideal system without any unbalance, both the journal center and central disk center quickly reach a steady state at the running speed of 8000 rpm. Due to the existence of shaft unbalance, the steady states of both the journal center and the

Table 6.9 Specification of a sample rotor-bearing system

Journal diameter, D	25.4 mm
Length of journal bearing, L	12.7 mm
Mass of the rotor, $2m$	5.4523 kg
Radial clearance, C	0.0508 mm
Shaft unbalance, $2mu$	0.000004 kg·m
Lubricant grade	ISO 32
Inlet temperature range, T_{in}	40°C
Rotating speed	8000 rpm

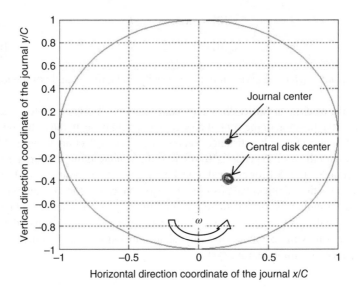

Figure 6.35   Orbits of journal and central disk

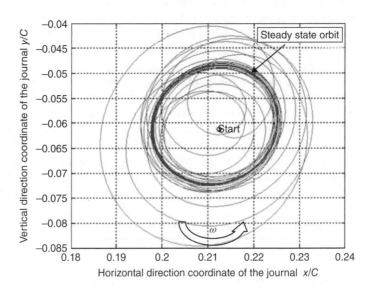

Figure 6.36   Zoom-in view of the journal orbit

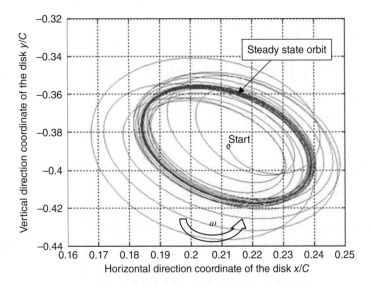

Figure 6.37   Zoom-in view of the central disk orbit

central disk center are in a *synchronous* whirl orbit. An interesting phenomenon is that the long axis of the central disk orbit has a phase lag of about 90° from the long axis of the journal orbit. This is caused by the mass inertia of the central disk.

6.6   Turbulence Effects

So far, all the applications we have presented pertain to a rotor-bearing system that operates in the laminar regime. This section shows that the steady state and dynamic characteristics of a rotor-bearing system with turbulent effects can be conveniently evaluated by applying the Hopf bifurcation theory to the equations of motion. Turbulence effects on the stiffness and damping coefficients and stability characteristics of a rotor-bearing system are also discussed.

6.6.1   Governing Equations for Turbulent Flow

The equations of motion of rotor-bearing systems for both turbulent and laminar flows are presented in Chapter 2. For convenience of further analysis, the equations of motion for short bearings (Eqs. 2.7–2.10) are repeated here for the application of Hopf bifurcation theory (Wang and Khonsari, 2006a).

$$\dot{x}_1 = x_2 \qquad\qquad (6.95)$$

$$\dot{x}_2 = x_1 x_4^2 + \frac{f_\varepsilon}{mC\omega^2} + \frac{g}{C\omega^2}\cos x_3$$

$$= x_1 x_4^2 - \frac{\Gamma}{\bar{\omega}}\left\{(1-2x_4)\left[\frac{2(a_z-2b_z)x_1^2+2b_z}{\left(1-x_1^2\right)^2} - \frac{b_z}{x_1}\ln\left(\frac{1+x_1}{1-x_1}\right)\right]\right.$$

$$\left. + \pi x_2\left[\frac{a_z+5b_z+2(a_z-3b_z)x_1^2-\left(2b_z/x_1^2\right)}{\left(1-x_1^2\right)^{5/2}} + \frac{2b_z}{x_1^2}\right]\right\} + \frac{1}{\bar{\omega}^2}\cos x_3 \tag{6.96}$$

$$\dot{x}_3 = x_4 \tag{6.97}$$

$$\dot{x}_4 = -\frac{2x_2 x_4}{x_1} + \frac{f_\phi}{x_1 mC\omega^2} - \frac{g}{C\omega^2 x_1}\sin x_3$$

$$= -\frac{2x_2 x_4}{x_1} + \frac{\Gamma}{\bar{\omega}x_1^2}\left\{\pi(1-2x_4)\left[\frac{2b_z+(a_z-3b_z)x_1^2}{2\left(1-x_1^2\right)^{3/2}} - b_z\right]\right.$$

$$\left. + 4x_2\left[\frac{b_z+(a_z-2b_z)x_1^2}{\left(1-x_1^2\right)^2} - \frac{b_z}{2x_1}\ln\left(\frac{1+x_1}{1-x_1}\right)\right]\right\} - \frac{1}{\bar{\omega}^2 x_1}\sin x_3 \tag{6.98}$$

where $x_1 = \varepsilon$, $x_2 = \dot{\varepsilon}$, $x_3 = \phi$, $x_4 = \dot{\phi}$, $\Gamma = \mu R L^3 / \left(2mC^{2.5}g^{0.5}\right)$, and $\bar{\omega} = \omega\sqrt{C/g}$. g is the gravitational constant. Γ is the dimensionless system characteristic number. The above system of Equations 6.95–6.98 is of the form

$$\dot{\mathbf{x}} = \mathbf{f}(\mathbf{x},\bar{\omega}) \tag{6.99}$$

The steady-state equilibrium position $\mathbf{x}_s$ in terms of $(x_{1s} = \varepsilon_s$, $x_{2s} = \dot{\varepsilon}_s$, $x_{3s} = \phi_s$, $x_{4s} = \dot{\phi}_s)$ is given by Equation 2.12.

Since the equation of motion (Eq. 6.99) has the suitable form of $\dot{\mathbf{x}} = \mathbf{f}(\mathbf{x},\bar{\omega})$ and possesses a steady-state equilibrium position $\mathbf{x}_s$ (the stationary point as required by assumption (i) in Section 4.1), the Hopf bifurcation theory can be applied to analyze its behavior.

With the application of the Hopf bifurcation theory described in Chapter 4, the right-hand side of the nonlinear equations of motion (6.99) is expanded in a Taylor series about the steady-state equilibrium position ($\mathbf{x} = \mathbf{x}_s$) as follows.

$$\mathbf{f}(\mathbf{x},\bar{\omega}) = \mathbf{f}(\mathbf{x}_s,\bar{\omega}) + \frac{\partial \mathbf{f}}{\partial \mathbf{x}}(\mathbf{x}_s,\bar{\omega})\Delta\mathbf{x} + \frac{\partial^2 \mathbf{f}}{\partial \mathbf{x}^2}(\mathbf{x}_s,\bar{\omega})(\Delta\mathbf{x})^2 + \frac{\partial^3 \mathbf{f}}{\partial \mathbf{x}^3}(\mathbf{x}_s,\bar{\omega})(\Delta\mathbf{x})^3 + \text{H.O.T.} \tag{6.100}$$

where $\Delta\mathbf{x}(\bar{\omega}) = \mathbf{x}(\bar{\omega}) - \mathbf{x}_s(\bar{\omega})$ and H. O. T. represent the higher order terms in the Taylor series expansion. The zeroth-order term $\mathbf{f}(\mathbf{x}_s,\bar{\omega})$ is used to determine the steady state equilibrium position. The first-order term $\partial \mathbf{f}/\partial \mathbf{x}(\mathbf{x}_s, \bar{\omega})$—often referred to as the Jacobian matrix of the equations of motion (6.99)—is used to determine the dynamic performance through analysis of the eigenvalues. The second-order term $\partial^2 \mathbf{f}/\partial \mathbf{x}^2 (\mathbf{x}_s, \bar{\omega})$ and third-order term $\partial^3 \mathbf{f}/\partial \mathbf{x}^3 (\mathbf{x}_s, \bar{\omega})$ are used to determine the stability, amplitude, and frequency of the periodic solutions.

The steady-state equilibrium position $\mathbf{x}_s$ in terms of $(x_{1s} = \varepsilon_s,\ x_{2s} = \dot{\varepsilon}_s,\ x_{3s} = \phi_s,\ x_{4s} = \dot{\phi}_s)$ can be found analytically by letting the zeroth term $\mathbf{f}(\mathbf{x}_s, \bar{\omega}) = 0$. The subscript s denotes the steady-state equilibrium position. The resulting equations for the steady state equilibrium position of the system with consideration of the turbulent effects are (Wang and Khonsari, 2006a)

$$\left[\frac{2(a_z - 2b_z)x_{1s}^2 + 2b_z}{\left(1 - x_{1s}^2\right)^2} - \frac{b_z}{x_{1s}}\ln\left(\frac{1 + x_{1s}}{1 - x_{1s}}\right)\right]^2 + \frac{\pi^2}{x_{1s}^2}\left[\frac{2b_z + (a_z - 3b_z)x_{1s}^2}{2\left(1 - x_{1s}^2\right)^{\frac{3}{2}}} - b_z\right]^2 = \left(\frac{1}{\Gamma\bar{\omega}}\right)^2$$

$$x_{2s} = 0$$

$$x_{3s} = \tan^{-1}\left(\frac{\dfrac{\pi}{2}\left[2b_z + (a_z - 3b_z)x_{1s}^2\right]\left(1 - x_{1s}^2\right)^{\frac{1}{2}} - \pi b_z\left(1 - x_{1s}^2\right)^2}{\left[2(a_z - 2b_z)x_{1s}^2 + 2b_z\right]x_{1s} - b_z\left(1 - x_{1s}^2\right)^2\ln\left(\dfrac{1 + x_{1s}}{1 - x_{1s}}\right)}\right)$$

$$x_{4s} = 0 \tag{6.101}$$

Since $\Gamma\bar{\omega} = 2S\pi(L/D)^2$, the first equation in Equation 6.101 describes a relationship between the Sommerfeld number S and the eccentricity, ε_s, and the third equation gives an expression for the attitude angle, ϕ_s in terms of the steady state eccentricity ε_s.

Under the laminar flow condition ($a_z = 1$ and $b_z = 0$), the simplified Equation 6.101 for the steady state equilibrium position of the system is (Wang and Khonsari, 2006a)

$$\frac{x_{1s}\sqrt{16x_{1s}^2 + \pi^2\left(1 - x_{1s}^2\right)}}{\left(1 - x_{1s}^2\right)^2} = \frac{1}{S\pi(L/D)^2}$$

$$x_{2s} = 0$$

$$x_{3s} = \tan^{-1}\left(\frac{\pi\sqrt{1 - x_{1s}^2}}{4x_{1s}}\right) \tag{6.102}$$

$$x_{4s} = 0$$

These equations are consistent with those obtained by direct solution of the Reynolds equation based on the infinitely short bearing assumption and laminar flow condition in Chapter 1, and the steady-state equilibrium position can be rewritten as $\mathbf{x}_s = (\varepsilon_s, 0, \phi_s, 0)$.

The critical speed $\bar{\omega}_{st}$ beyond which the system becomes unstable can be identified as the speed at which a pair of the eigenvalues of the Jacobian matrix $\mathbf{f}_\mathbf{x}(\mathbf{x}_s, \bar{\omega})$ are equal to $\pm i\beta(\bar{\omega}_{st})$ while the real parts of all the other eigenvalues are purely negative. The result of the threshold speed $\bar{\omega}_{st}$ is identical to that obtained by the traditionally linearized stability analysis in Chapter 3 since it is also based on the same perturbation of the fluid force $\mathbf{f}(\mathbf{x}, \bar{\omega}) \approx \mathbf{f}(\mathbf{x}_s, \bar{\omega}) + \partial \mathbf{f}/\partial \mathbf{x}(\mathbf{x}_s, \bar{\omega})\Delta\mathbf{x}$, where $\mathbf{f}(\mathbf{x}_s, \bar{\omega}) = 0$.

Equation (6.103) shows that the derived Jacobian matrix of the equations of motion (6.99) at the stationary point $\mathbf{x}_s$ where $\mathbf{f}(\mathbf{x}_s, \bar{\omega}) = 0$ is (Wang and Khonsari, 2006a)

$$
\mathbf{f}_\mathbf{x}(\mathbf{x}_s, \omega) = \frac{\partial \mathbf{f}}{\partial \mathbf{x}}(\mathbf{x}_s, \omega)
$$

$$
= \begin{bmatrix}
0 & 1 & 0 & 0 \\[2mm]
\dfrac{1}{m\omega^2}\dfrac{\partial f_\varepsilon}{C\partial x_1} & \dfrac{1}{m\omega^2}\dfrac{\partial f_\varepsilon}{C\partial x_2} & -\dfrac{x_1}{m\omega^2}\left(-\dfrac{\partial f_\varepsilon}{Cx_1\partial x_3} + \dfrac{f_\phi}{Cx_1}\right) & \dfrac{x_1}{m\omega^2}\dfrac{\partial f_\varepsilon}{Cx_1\partial x_4} \\[2mm]
0 & 0 & 0 & 1 \\[2mm]
\dfrac{1}{x_1 m\omega^2}\dfrac{\partial f_\phi}{C\partial x_1} & \dfrac{1}{x_1 m\omega^2}\dfrac{\partial f_\phi}{C\partial x_2} & \dfrac{1}{m\omega^2}\left(\dfrac{\partial f_\phi}{Cx_1\partial x_3} + \dfrac{f_\varepsilon}{Cx_1}\right) & \dfrac{1}{m\omega^2}\dfrac{\partial f_\phi}{Cx_1\partial x_4}
\end{bmatrix}
$$

$$(6.103)$$

The definitions of the linearized stiffness coefficients k_{ij} $(i, j = \varepsilon, \phi)$ and the damping coefficients b_{ij} $(i, j = \varepsilon, \phi)$ are given in Section 3.1 by Equation 3.18. For convenience, Equation 3.18 is rewritten as below:

$$
\begin{bmatrix} k_{\varepsilon\varepsilon} & k_{\varepsilon\phi} \\ k_{\phi\varepsilon} & k_{\phi\phi} \end{bmatrix} = \begin{bmatrix} -\dfrac{\partial f_\varepsilon}{C\partial \varepsilon} & -\dfrac{\partial f_\varepsilon}{C\varepsilon\partial \phi} + \dfrac{f_\phi}{C\varepsilon} \\[3mm] -\dfrac{\partial f_\phi}{C\partial \varepsilon} & -\dfrac{\partial f_\phi}{C\varepsilon\partial \phi} - \dfrac{f_\varepsilon}{C\varepsilon} \end{bmatrix} \quad \text{and} \quad \begin{bmatrix} b_{\varepsilon\varepsilon} & b_{\varepsilon\phi} \\ b_{\phi\varepsilon} & b_{\phi\phi} \end{bmatrix} = \begin{bmatrix} -\dfrac{\partial f_\varepsilon}{C\partial \dot{\varepsilon}} & -\dfrac{\partial f_\varepsilon}{C\varepsilon\partial \dot{\phi}} \\[3mm] -\dfrac{\partial f_\phi}{C\partial \dot{\varepsilon}} & -\dfrac{\partial f_\phi}{C\varepsilon\partial \dot{\phi}} \end{bmatrix}
$$

$$(6.104)$$

where all the matrix elements are evaluated at the equilibrium position $\mathbf{x}_s = (\varepsilon_s, 0, \phi_s, 0)$ and "." represents d/dt.

In terms of these linearized stiffness and damping coefficients given by Equations 6.104, letting $\bar{t} = \omega t$, $x_1 = \varepsilon$, $x_2 = \dot{\varepsilon}$, $x_3 = \phi$ and $x_4 = \dot{\phi}$, the Jacobian matrix—Equation 6.103—can be rewritten as (Wang and Khonsari, 2006a)

$$\mathbf{f_x}(\mathbf{x_s},\omega) = \begin{bmatrix} 0 & 1 & 0 & 0 \\ -\dfrac{1}{m\omega^2}k_{\varepsilon\varepsilon} & -\dfrac{1}{m\omega}b_{\varepsilon\varepsilon} & -\dfrac{\varepsilon}{m\omega^2}k_{\varepsilon\phi} & -\dfrac{\varepsilon}{m\omega}b_{\varepsilon\phi} \\ 0 & 0 & 0 & 1 \\ -\dfrac{1}{\varepsilon m\omega^2}k_{\phi\varepsilon} & -\dfrac{1}{\varepsilon m\omega}b_{\phi\varepsilon} & -\dfrac{1}{m\omega^2}k_{\phi\phi} & -\dfrac{1}{m\omega}b_{\phi\phi} \end{bmatrix} \qquad (6.105)$$

Substituting the fluid force component expressions (Eqs. 1.21 and 1.22) in Section 1.2 into Equation 6.104 and simplifying the resulting expressions, it can be shown that the elements of the linearized stiffness matrix k_{ij} $(i, j = \varepsilon, \phi)$ for turbulent flow are (Wang and Khonsari, 2006a)

$$k_{\varepsilon\varepsilon} = \frac{\omega\mu R L^3}{2C^3}\left[\frac{4(a_z - 2b_z)\varepsilon(1+\varepsilon^2) + 8b_z\varepsilon}{(1-\varepsilon^2)^3} + \frac{b_z}{\varepsilon^2}\ln\left(\frac{1+\varepsilon}{1-\varepsilon}\right) - \frac{2b_z}{\varepsilon(1-\varepsilon^2)}\right]$$

$$k_{\varepsilon\phi} = \frac{\pi\omega\mu R L^3}{2C^3\varepsilon^2}\left[\frac{2b_z + (a_z - 3b_z)\varepsilon^2}{2(1-\varepsilon^2)^{3/2}} - b_z\right]$$

$$k_{\phi\varepsilon} = -\frac{\pi\omega\mu R L^3}{4C^3\varepsilon^2(1-\varepsilon^2)^{5/2}}\left[(a_z - 3b_z)(1+2\varepsilon^2)\varepsilon^2 + 2b_z(1-\varepsilon^2)^{5/2} + 2b_z(4\varepsilon^2 - 1)\right]$$

$$k_{\phi\phi} = \frac{\omega\mu R L^3}{2C^3\varepsilon}\left[\frac{2(a_z - 2b_z)\varepsilon^2 + 2b_z}{(1-\varepsilon^2)^2} - \frac{b_z}{\varepsilon}\ln\left(\frac{1+\varepsilon}{1-\varepsilon}\right)\right]$$

$$(6.106)$$

The elements of the damping matrix b_{ij} $(i, j = \varepsilon, \phi)$ for turbulent flow are (Wang and Khonsari, 2006a)

$$b_{\varepsilon\varepsilon} = \frac{\pi\mu R L^3}{2C^3}\left[\frac{a_z + 5b_z + 2(a_z - 3b_z)\varepsilon^2 - \dfrac{2b_z}{\varepsilon^2}}{(1-\varepsilon^2)^{5/2}} + \frac{2b_z}{\varepsilon^2}\right]$$

$$b_{\varepsilon\phi} = -\frac{2\mu R L^3}{C^3\varepsilon}\left[\frac{(a_z - 2b_z)\varepsilon^2 + b_z}{(1-\varepsilon^2)^2} - \frac{b_z}{2\varepsilon}\ln\left(\frac{1+\varepsilon}{1-\varepsilon}\right)\right]$$

$$(6.107)$$

$$b_{\phi\varepsilon} = -\frac{2\mu R L^3}{C^3\varepsilon}\left[\frac{(a_z - 2b_z)\varepsilon^2 + b_z}{(1-\varepsilon^2)^2} - \frac{b_z}{2\varepsilon}\ln\left(\frac{1+\varepsilon}{1-\varepsilon}\right)\right]$$

$$b_{\phi\phi} = \frac{\pi\mu R L^3}{C^3\varepsilon^2}\left[\frac{2b_z + (a_z - 3b_z)\varepsilon^2}{2(1-\varepsilon^2)^{3/2}} - b_z\right]$$

Simplifying Equations 6.106 and 6.107 with $a_z = 1$ and $b_z = 0$ can yield the equations of the stiffness and damping coefficients for laminar fluid flow, which are consistent with Equations 3.21 and 3.22 derived in Chapter 3.

Letting $\bar{k}_{ij} = (C/R)^3/(\mu\omega L)k_{ij}$, $(i,j=\varepsilon,\phi)$ and $\bar{b}_{ij} = (C/R)^3/(\mu L)b_{ij}$, $(i,j=\varepsilon,\phi)$, the normalized stiffness coefficients $\bar{k}_{ij}$ $(i,j=\varepsilon,\phi)$ become (Wang and Khonsari, 2006a)

$$\bar{k}_{\varepsilon\varepsilon} = \left(\frac{L}{D}\right)^2 2\left[\frac{4(a_z-2b_z)\varepsilon(1+\varepsilon^2)+8b_z\varepsilon}{(1-\varepsilon^2)^3} + \frac{b_z}{\varepsilon^2}\ln\left(\frac{1+\varepsilon}{1-\varepsilon}\right) - \frac{2b_z}{\varepsilon(1-\varepsilon^2)}\right]$$

$$\bar{k}_{\varepsilon\phi} = \left(\frac{L}{D}\right)^2 \frac{2\pi}{\varepsilon^2}\left[\frac{2b_z+(a_z-3b_z)\varepsilon^2}{2(1-\varepsilon^2)^{3/2}} - b_z\right]$$

$$\bar{k}_{\phi\varepsilon} = -\left(\frac{L}{D}\right)^2 \frac{\pi\left[(a_z-3b_z)(1+2\varepsilon^2)\varepsilon^2+2b_z(1-\varepsilon^2)^{5/2}+2b_z(4\varepsilon^2-1)\right]}{\varepsilon^2(1-\varepsilon^2)^{5/2}} \qquad (6.108)$$

$$\bar{k}_{\phi\phi} = \left(\frac{L}{D}\right)^2 \frac{2}{\varepsilon}\left[\frac{2(a_z-2b_z)\varepsilon^2+2b_z}{(1-\varepsilon^2)^2} - \frac{b_z}{\varepsilon}\ln\left(\frac{1+\varepsilon}{1-\varepsilon}\right)\right]$$

The normalized damping coefficients $\bar{b}_{ij}$ $(i,j=\varepsilon,\phi)$ are (Wang and Khonsari, 2006a)

$$\bar{b}_{\varepsilon\varepsilon} = \left(\frac{L}{D}\right)^2 2\pi\left[\frac{a_z+5b_z+2(a_z-3b_z)\varepsilon^2-\frac{2b_z}{\varepsilon^2}}{(1-\varepsilon^2)^{5/2}} + \frac{2b_z}{\varepsilon^2}\right]$$

$$\bar{b}_{\varepsilon\phi} = -\left(\frac{L}{D}\right)^2 \frac{8}{\varepsilon}\left[\frac{(a_z-2b_z)\varepsilon^2+b_z}{(1-\varepsilon^2)^2} - \frac{b_z}{2\varepsilon}\ln\left(\frac{1+\varepsilon}{1-\varepsilon}\right)\right]$$

$$\bar{b}_{\phi\varepsilon} = -\left(\frac{L}{D}\right)^2 \frac{8}{\varepsilon}\left[\frac{b_z+(a_z-2b_z)\varepsilon^2}{(1-\varepsilon^2)^2} - \frac{b_z}{2\varepsilon}\ln\left(\frac{1+\varepsilon}{1-\varepsilon}\right)\right] \qquad (6.109)$$

$$\bar{b}_{\phi\phi} = \left(\frac{L}{D}\right)^2 \frac{4\pi}{\varepsilon^2}\left[\frac{2b_z+(a_z-3b_z)\varepsilon^2}{2(1-\varepsilon^2)^{3/2}} - b_z\right]$$

In Section 3.1.2, letting $\bar{K}_{ij}=\pi S\bar{k}_{ij}$, $\bar{B}_{ij}=\pi S\bar{b}_{ij}$ $(i,j=\varepsilon,\phi)$, we derived the instability threshold speed $\bar{\omega}_{st}$ based on the traditionally linearized stiffness and damping coefficients. The definition of the threshold speed $\bar{\omega}_{st}$ defined in HBT is identical to that obtained by the traditionally linearized stability analysis in Chapter 3 (Wang and Khonsari, 2006a). Pertinent Equations 3.33–3.35, rewritten below for convenience, are utilized to calculate $\bar{\omega}_{st}$, $\bar{\omega}_s$, and then Ω. $\bar{\omega}_s$ is the dimensionless whirl speed and $\Omega=\bar{\omega}_s/\bar{\omega}_{st}$ is the whirl frequency ratio at the threshold speed $\bar{\omega}_{st}$.

$$\bar{\omega}_{\text{st}} = \bar{\omega}_{\text{s}} \sqrt{\frac{\bar{B}_{\varepsilon\varepsilon}\bar{B}_{\phi\phi} - \bar{B}_{\varepsilon\phi}\bar{B}_{\phi\varepsilon}}{\left(\bar{\omega}_{\text{s}}^2 - \bar{K}_{\varepsilon\varepsilon}\right)\left(\bar{\omega}_{\text{s}}^2 - \bar{K}_{\phi\phi}\right) - \bar{K}_{\varepsilon\phi}\bar{K}_{\phi\varepsilon}}} \qquad (6.110)$$

$$\bar{\omega}_{\text{s}} = \sqrt{\frac{\bar{K}_{\varepsilon\varepsilon}\bar{B}_{\phi\phi} + \bar{K}_{\phi\phi}\bar{B}_{\varepsilon\varepsilon} - \bar{K}_{\varepsilon\phi}\bar{B}_{\phi\varepsilon} - \bar{K}_{\phi\varepsilon}\bar{B}_{\varepsilon\phi}}{\bar{B}_{\varepsilon\varepsilon} + \bar{B}_{\phi\phi}}} \qquad (6.111)$$

$$\Omega^2 = \frac{\bar{\omega}_{\text{s}}^2}{\bar{\omega}_{\text{st}}^2} = \frac{\left(\bar{\omega}_{\text{s}}^2 - \bar{K}_{\varepsilon\varepsilon}\right)\left(\bar{\omega}_{\text{s}}^2 - \bar{K}_{\phi\phi}\right) - \bar{K}_{\varepsilon\phi}\bar{K}_{\phi\varepsilon}}{\bar{B}_{\varepsilon\varepsilon}\bar{B}_{\phi\phi} - \bar{B}_{\varepsilon\phi}\bar{B}_{\phi\varepsilon}} \qquad (6.112)$$

6.6.2 Effects of Turbulence on the Dynamic Performance

Based on the analytical expressions derived in the previous section, Figures 6.38, 6.39, 6.40, 6.41, and 6.42 are drawn to show how turbulence affects the eccentricity ratio, the normalized stiffness coefficients $\bar{k}_{ij}$ ($i, j = \varepsilon, \phi$) and damping coefficients $\bar{b}_{ij}$ ($i, j = \varepsilon, \phi$), the threshold speed $\bar{\omega}_{\text{st}}$, and the whirl frequency ratio Ω. Here, a rotor-bearing system with $L/D = 0.5$ is analyzed.

Figure 6.38 shows that turbulence has a remarkable influence on the steady-state eccentricity ratio when $0.05 \leq S \leq 4$. For example, when $S = 0.4$, the eccentricity ratio $\varepsilon = 0.51$ if the flow is laminar. At the same Sommerfeld number, the eccentricity ratio is shown to drop to $\varepsilon = 0.42$ with Re = 2500 and to $\varepsilon = 0.36$ with Re = 5000 and further down to $\varepsilon = 0.27$ with Re = 10 000. However, the effect of turbulence on the eccentricity ratio fades out as Sommerfeld number S becomes greater than 4 or less than 0.05.

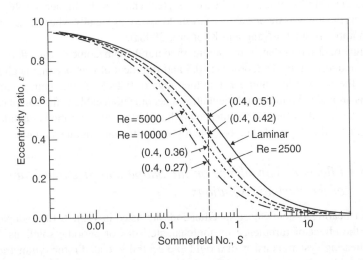

Figure 6.38 Turbulent effects on the eccentricity ratio. From Wang and Khonsari (2006a) © Elsevier Limited.

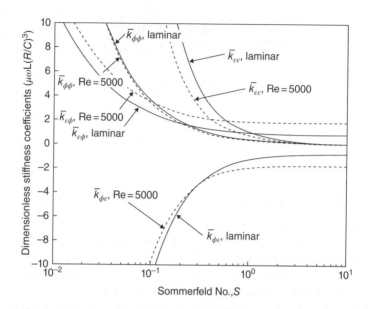

Figure 6.39   Turbulent effects on the dimensionless stiffness coefficients $\bar{k}_{ij}$ $(i, j = \varepsilon, \phi)$. From Wang and Khonsari (2006a) © Elsevier Limited.

Figures 6.39 and 6.40 show that turbulence has a significant influence on both stiffness and damping coefficients of the rotor-bearing system.

Figure 6.41 shows that the effect of turbulence on the threshold speed $\bar{\omega}_{st}$ is consistent with that described by Hashimoto *et al.* (1987). Turbulence tends to deteriorate the stability of the rotor-bearing system and that the higher the Reynolds number, the lower is the instability threshold speed, especially when the Sommerfeld number is small (Wang and Khonsari, 2006a).

Figure 6.42 shows that, as the Sommerfeld number S becomes greater than 5, the whirl frequency ratio Ω approaches 0.5 and turbulent effects become negligible. When the Sommerfeld number is in the range of $0.2 \leq S \leq 5$, the whirl frequency ratio is around 0.5 and the turbulent effects become noticeable while complicated. When $S < 0.2$, the effect of turbulence on the whirl frequency ratio is remarkable. The greater the Reynolds number, the higher is the whirl frequency ratio.

6.6.3   Effects of Turbulence on the Shape and Size and Stability of the Periodic Solutions

A rotor-bearing system whose specifications are listed in Table 6.10 is adopted to show the effect of turbulence on the characteristics of periodic solutions. This rotor-bearing system consists of a rigid and centrally loaded rotor symmetrically supported by two identical fluid-film journal bearings.

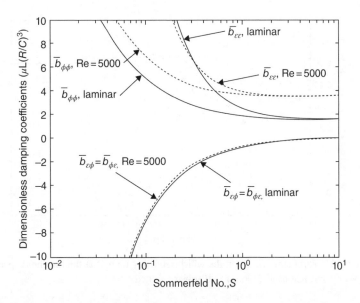

Figure 6.40   Turbulent effects on the dimensionless damping coefficients $\bar{b}_{ij}$ ($i,j=\varepsilon,\phi$). From Wang and Khonsari (2006a) © Elsevier Limited.

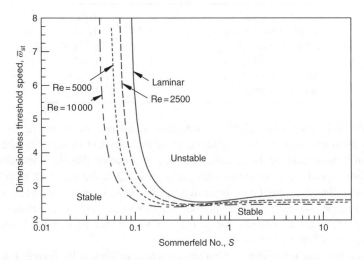

Figure 6.41   Turbulent effects on the dimensionless threshold speed $\bar{\omega}_{st}$. From Wang and Khonsari (2006a) © Elsevier Limited.

Upon the application of Hopf bifurcation theory described in Chapter 4 on the equations of motion (Eq. 6.99) of the above rotor-bearing system, the Hopf bifurcation parameters shown in Table 6.11 are obtained.

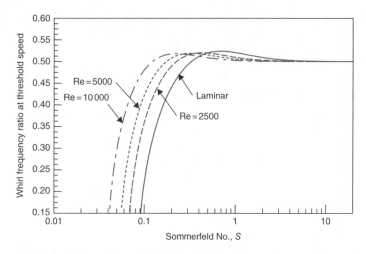

Figure 6.42   Turbulent effects on the whirl frequency ratio Ω at the threshold speed $\bar{\omega}_{st}$. From Wang and Khonsari (2006a) © Elsevier Limited.

Table 6.10   Specification of the rotor-bearing system (Wang and Khonsari, 2006a)

Journal diameter, D	0.076 m
Length of the bearing, L	0.038 m
Radial clearance, C	1.27×10^{-4} m
Mass of the rotor, $2m$	4.54 kg
Lubricant viscosity, μ	0.00138 Pa·s
Lubricant density, ρ	1000 kg/m^3

Table 6.11 shows the substantial differences between the results of the Hopf bifurcation parameters with consideration of the turbulent effects (Re = 2507 at $\bar{\omega}_{st}$) and those without the turbulent effects. It is shown that the dimensionless threshold speed $\bar{\omega}_{st}$ is dropped from 2.686 to 2.573 if the turbulent effects are included. Based on the theory presented in Chapter 4, since $\gamma_2 < 0$ and $\beta_2 > 0$ for both of these two cases, unstable periodic solutions (i.e., stability envelopes) exist for the system running speeds that are close to but less than the threshold speed $\bar{\omega}_{st}$ (Wang and Khonsari, 2006a).

From the Hopf bifurcation parameters presented in Table 6.11, Figures 6.43 and 6.44 are obtained using the Hopf bifurcation formulae (4.2–4.4) described in Section 4.2 to show the difference in the shapes and sizes of the periodic solutions with and without consideration of the turbulence effects.

Figure 6.43a shows the shape of the unstable periodic solution (stability envelope) as a function of system running speed close to but less than or equal to the critical speed $\bar{\omega}_{st}$ without consideration of turbulent effects. Each cross section of

Table 6.11 Hopf bifurcation parameters

	$\bar{\omega}_{st}$	γ_2	τ_2	β_2	$\beta(\bar{\omega}_{st})$	$\mathbf{v}_1$	$\mathbf{x}_s(\bar{\omega}_{st})$
Without consideration of turbulent effects	2.686	−4.050	0.289	0.688	0.509	$1.000 + 0.000i$	0.192
						$0.000 + 0.509i$	0
						$0.605 − 5.746i$	1.325
						$2.930 + 0.308i$	0
With consideration of turbulent effects (Re = 2507 at $\bar{\omega}_{st}$)	2.573	−9.599	0.108	1.374	0.503	$1.000 + 0.000i$	0.127
						$0.000 + 0.503i$	0
						$0.549 − 8.163i$	1.427
						$4.111 + 0.276i$	0

From Wang and Khonsari (2006a) © Elsevier Limited.

Figure 6.43a at a given dimensionless running speed $\bar{\omega}$ represents the unstable periodic solution (stability envelope) of the journal orbit at this specific running speed. Figure 6.43a shows that the periodic solutions shrink to a single point as the running speed approaches the critical value, $\bar{\omega}_{st}$. Figure 6.43b shows a similar profile but the results include turbulent effects (Re = 2507 at $\bar{\omega}_{st}$). To better understand the turbulent effects on the periodic solutions of journal orbit, Figure 6.44 is plotted to compare the bifurcation profiles with and without the turbulent effects.

Figure 6.44 compares two bifurcation profiles that depict the amplitudes of the periodic solutions corresponding to a set of system running speeds close to but less than or equal to the critical speed $\bar{\omega}_{st}$. The amplitude of the periodic solution corresponding to a given running speed $\bar{\omega}$ is bounded by $\varepsilon_s \pm \sqrt{(\bar{\omega} - \bar{\omega}_{st})/\gamma_2}$. It shrinks to one point as the running speed approaches the critical value $\bar{\omega}_{st}$. The threshold speed ($\bar{\omega}_{st} = 2.573$) with consideration of turbulent effects (Re = 2507 at $\bar{\omega}_{st}$) is significantly lower than that ($\bar{\omega}_{st} = 2.686$) without turbulent effects. If the threshold speeds are converted to a dimensional form, the threshold speed with consideration of turbulent effects will be 6829 rpm and the threshold speed without turbulent effects will rise to 7130 rpm.

In Figure 6.45, the system operating line is drawn using the expression $\bar{\omega} = \pi m g^{0.5} C^{2.5}/(\mu L R^3) S$, which describes the system's dimensionless running speed $\bar{\omega}$ in terms of the Sommerfeld number S. It is derived from the definition of the Sommerfeld number $S = \mu R L \omega/(\pi m g)(R/C)^2$ and $\bar{\omega} = \omega\sqrt{C/g}$. Figure 6.45 shows that 2.573 is the dimensionless threshold speed with consideration of the turbulent effects. With the turbulent effects neglected, the predicted dimensionless threshold speed is 2.686.

As an example, at the dimensionless running speed of $\bar{\omega} = 2.54$, Figure 6.46 compares the periodic solution with consideration of turbulent effects, i.e. a cross section of Figure 6.43b, and that without turbulent effects, i.e. a cross section of Figure 6.43a.

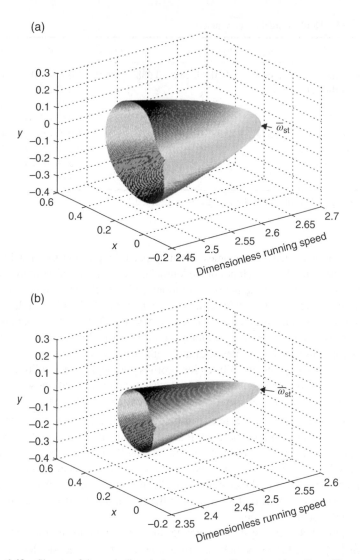

Figure 6.43 Shapes of the periodic solutions corresponding to a series of specific running speeds. (a) Results neglecting turbulent effects. (b) Results with consideration of the turbulent effects (Re = 2507 at $\bar{\omega}_{st}$). From Wang and Khonsari (2006a) © Elsevier Limited.

Figure 6.46 shows that the periodic solution with consideration of turbulent effects is much smaller than that without the turbulent effects when the dimensionless running speed $\bar{\omega} = 2.54$. According to the definition of the unstable periodic solution in Chapter 4 and its correlation with the stability envelope established in Chapter 5, if the journal is released from a position inside the unstable periodic

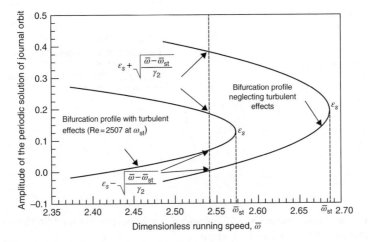

Figure 6.44 Bifurcation profiles with and without consideration of turbulence effects. From Wang and Khonsari (2006a) © Elsevier Limited.

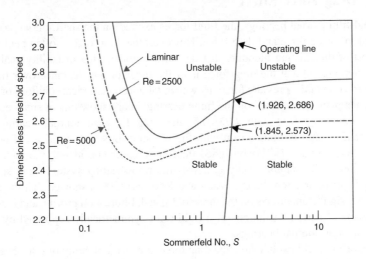

Figure 6.45 Prediction of the threshold speed using Figure 6.41. From Wang and Khonsari (2006a) © Elsevier Limited.

solution (i.e., stability envelope), the system tends to asymptotically approach the steady-state equilibrium position; if the journal is released from a position outside the unstable periodic solution (stability envelope), the orbit of the journal tends to asymptotically approach the clearance circle, in other words, it becomes unstable (Wang and Khonsari, 2006c). Therefore, Figure 6.46 shows that the turbulent effects significantly reduce the envelope of stability corresponding to the same system's running speed (Wang and Khonsari, 2006a).

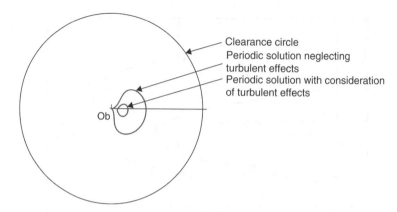

Figure 6.46   Periodic solutions with and without turbulent effects at $\bar{\omega} = 2.54$. From Wang and Khonsari (2006a) © Elsevier Limited.

6.7   Drag Force Effect

In fluid-film journal bearings, the fluid forces exerted on the shaft journal consist of both pressure force and drag force. However, the drag force is often neglected in most of the existing literature that deal with rotor bearing instability analysis. It is often argued that the drag force is of the order of the clearance over radius (C/R) fraction of the pressure force. However, tabulated numerical solutions of performance parameters based on the finite bearing theory show that at small eccentricity ratios ($\varepsilon \approx 0.1$) with $L/D \leq 0.5$, the drag force in journal bearings is considerably high in comparison with the pressure force (Khonsari and Booser, 2008). Akers *et al.* (1971) concluded that, in all cases considered, the inclusion of friction in the stability analysis makes the rotor-bearing system more stable. Using Hopf bifurcation theory, Wang and Khonsari (2005) showed that the drag force has significant effects on the threshold speed, bifurcation profile, and the size and shape of periodic solutions of a rigid rotor symmetrically supported by two identical short journal bearings.

This section will explain how the drag force affects the stability of rotor-bearing system. To this end, the Hopf bifurcation theory (HBT) presented in Chapter 4 is utilized to examine the dynamic behavior of the system by analyzing the equations of motion directly.

6.7.1   Dynamic Fluid Forces in Journal Bearings

Figure 6.47 shows the radial and tangential components of the pressure and drag forces applied on the shaft journal.

Referring to Chapter 1, assuming constant oil viscosity throughout the fluid film and based on the short bearing theory with half-Sommerfeld boundary conditions,

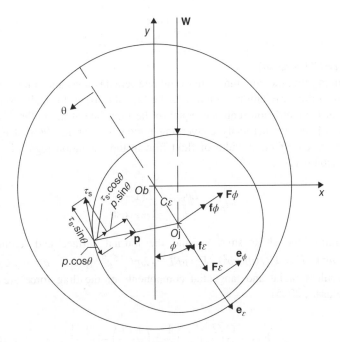

Figure 6.47   Radial and tangential components of the fluid forces exerted on the journal. From Wang and Khonsari (2005) © SAGE Publications Ltd.

we proceed by deriving the dimensionless form of the pressure force components by integrating the fluid pressure distribution obtained from the solution to the Reynolds equation. For further analysis, Equations 1.25 and 1.26 are rewritten in the following form as Equations 6.113 and 6.114.

$$\bar{f}_\varepsilon = -\frac{\Gamma}{\bar{\omega}}\left[\frac{2\varepsilon^2\left(1-2\dot\phi\right)}{\left(1-\varepsilon^2\right)^2} + \frac{\pi\dot\varepsilon(1+2\varepsilon^2)}{\left(1-\varepsilon^2\right)^{5/2}}\right] \tag{6.113}$$

$$\bar{f}_\phi = \frac{\Gamma}{\bar{\omega}}\left[\frac{\pi\left(1-2\dot\phi\right)\varepsilon}{2\left(1-\varepsilon^2\right)^{3/2}} + \frac{4\dot\varepsilon\varepsilon}{\left(1-\varepsilon^2\right)^2}\right] \tag{6.114}$$

where "." represents $d/(\omega dt)$, $\Gamma = \mu R L^3 / \left(2mC^{2.5}g^{0.5}\right)$, and $\bar{\omega} = \omega\sqrt{C/g}$.

Equation 6.115 gives the viscous shear stress at the journal surface for short bearing (Dubois and Ocvirk, 1953).

$$\tau_s = \frac{\mu \omega R}{h} \qquad (6.115)$$

where $h = C(1 + \varepsilon \cos\theta)$.

Assuming that the circumferential flow rate remains constant in the geometrically diverging cavitation region $(\pi \leq \theta < 2\pi)$, the flow through each cross section of the cavitation region is equal to the flow rate at the outlet $(\theta = \pi$ and $h = h_{min})$ of the geometrically converging region (Cameron, 1966). Under this assumption, an effective width of fluid film in the cavitation region is defined by Equation 6.116.

$$L_{eff} = \frac{L h_{min}}{h} = \frac{L(1-\varepsilon)}{1 + \varepsilon \cos\theta} \qquad (6.116)$$

Integrating the shear stress around the journal surface and letting $\bar{F}_\varepsilon = F_\varepsilon / (mC\omega^2)$, $\bar{F}_\phi = F_\phi / (mC\omega^2)$, $\Gamma = \mu R L^3 / (2mC^{2.5}g^{0.5})$, and $\bar{\omega} = \omega\sqrt{C/g}$, the dimensionless radial and tangential components of the drag force are (Wang and Khonsari, 2005)

$$\bar{F}_\varepsilon = -\frac{C}{R}\left(\frac{D}{L}\right)^2 \frac{\Gamma}{\bar{\omega}}\left[\frac{1}{2\varepsilon}\ln\left(\frac{1+\varepsilon}{1-\varepsilon}\right) - \frac{1}{1+\varepsilon}\right] \qquad (6.117)$$

$$\bar{F}_\phi = \frac{C}{R}\left(\frac{D}{L}\right)^2 \frac{\Gamma}{\bar{\omega}}\frac{\pi}{2\varepsilon}\left[1 - \frac{1-\varepsilon^3}{\left(1-\varepsilon^2\right)^{3/2}}\right] \qquad (6.118)$$

6.7.2   Equations of Motion

The dimensionless equations of motion for a rotor-bearing system consisting of a rigid and perfectly balanced rotor (with mass 2m) symmetrically supported by two identical fluid-film journal bearings are (Mitchell *et al.*, 1965; Wang and Khonsari, 2005)

$$\ddot{\varepsilon} - \varepsilon\dot{\phi}^2 - \bar{f}_\varepsilon - \bar{F}_\varepsilon - \frac{1}{\bar{\omega}^2}\cos\phi = 0 \qquad (6.119)$$

$$\ddot{\phi} + \frac{2\dot{\varepsilon}\dot{\phi}}{\varepsilon} - \frac{\bar{f}_\phi}{\varepsilon} - \frac{\bar{F}_\phi}{\varepsilon} + \frac{1}{\bar{\omega}^2\varepsilon}\sin\phi = 0 \qquad (6.120)$$

where "." represents $d/(\omega dt)$. $\bar{f}_\varepsilon$ and $\bar{f}_\phi$ are the dimensionless radial and tangential pressure force components as given by Equations 6.113 and 6.114; $\bar{F}_\varepsilon$ and $\bar{F}_\phi$ are the dimensionless radial and tangential drag force components as given by Equations 6.117 and 6.118.

To solve the equations of motion (Eqs. 6.119 and 6.120), these two second-order nonlinear equations are first decomposed into four first-order equations. Let $x_1 = \varepsilon$, $x_2 = \dot{\varepsilon}$, $x_3 = \phi$, and $x_4 = \dot{\phi}$; then the decomposed equations of motion become (Wang and Khonsari, 2005)

$$\dot{x}_1 = x_2 \tag{6.121}$$

$$\dot{x}_2 = x_1 x_4^2 - \frac{\Gamma}{\bar{\omega}} \left[\frac{2x_1^2(1 - 2x_4)}{\left(1 - x_1^2\right)^2} + \frac{\pi x_2\left(1 + 2x_1^2\right)}{\left(1 - x_1^2\right)^{5/2}} \right] + \frac{1}{\bar{\omega}^2}\cos x_3$$

$$- \frac{C\Gamma}{R\bar{\omega}}\left(\frac{D}{L}\right)^2 \left[\frac{1}{2x_1}\ln\left(\frac{1 + x_1}{1 - x_1}\right) - \frac{1}{1 + x_1} \right] \tag{6.122}$$

$$\dot{x}_3 = x_4 \tag{6.123}$$

$$\dot{x}_4 = -\frac{2x_2 x_4}{x_1} + \frac{\Gamma}{\bar{\omega}} \left[\frac{\pi(1 - 2x_4)}{2\left(1 - x_1^2\right)^{3/2}} + \frac{4x_2}{\left(1 - x_1^2\right)^2} \right] - \frac{1}{\bar{\omega}^2 x_1}\sin x_3$$

$$+ \frac{C\Gamma}{R\bar{\omega}}\left(\frac{D}{L}\right)^2 \frac{\pi}{2x_1} \left[1 - \frac{1 - x_1^3}{\left(1 - x_1^2\right)^{3/2}} \right] \tag{6.124}$$

The above system of equations (Eqs. 6.121–6.124) is of the form

$$\dot{\mathbf{x}} = \mathbf{f}(\mathbf{x}, \bar{\omega}) \tag{6.125}$$

which is suitable for the application of Hopf bifurcation theory described in Chapter 4. The steady state equilibrium position $\mathbf{x}_s$ in terms of ($x_{1s} = \varepsilon_s$, $x_{2s} = \dot{\varepsilon}_s$, $x_{3s} = \phi_s$, $x_{4s} = \dot{\phi}_s$) can be found analytically from letting $\mathbf{f}(\mathbf{x}_s, \bar{\omega}) = 0$. The subscript s denotes the steady state equilibrium position. The steady state equilibrium position can be rewritten as $\mathbf{x}_s = (\varepsilon_s, 0, \phi_s, 0)$.

6.7.3   Effects of Drag Force on the Hopf Bifurcation Profile

Consider the rotor-bearing system whose specifications are listed in Table 6.12. This rotor-bearing system consists of a rigid and perfectly balanced rotor symmetrically supported by two identical fluid-film journal bearings.

Based on the Hopf bifurcation theory presented in Chapter 4, the Hopf bifurcation parameters shown in Table 6.13 are obtained.

Table 6.13 shows the considerable differences between the results of the Hopf bifurcation parameters of the equations of motion (Eq. 6.125) with and without the

Table 6.12 Specifications of a rotor-bearing system (Wang and Khonsari, 2005)

Journal diameter, D	0.0254 m
Length of the bearing, L	0.0127 m
Radial clearance, C	50.8×10^{-6} m
Mass of the rotor, $2m$	5.4523 kg
Oil viscosity, μ	0.01 Pa·s

Table 6.13 Hopf bifurcation parameters

	$\bar{\omega}_{st}$	γ_2	τ_2	β_2	$\beta(\bar{\omega}_{st})$	$\mathbf{v}_1$	$\mathbf{x}_s(\bar{\omega}_{st})$
Without considera-tion of drag force	2.643	−1.754	0.396	0.308	0.515	$1.000 + 0.000i$	0.250
						$0.000 + 0.515i$	0
						$0.580 - 4.745i$	1.252
						$2.444 + 0.298i$	0
With consideration of drag force	2.689	−1.677	0.402	0.287	0.511	$1.000 + 0.000i$	0.247
						$0.000 + 0.511i$	0
						$0.575 - 4.776i$	1.247
						$2.443 + 0.294i$	0

From Wang and Khonsari (2005) © SAGE Publications Ltd.

consideration of the drag force effect. It is shown that the dimensionless instability threshold speed $\bar{\omega}_{st}$ increases from 2.643 to 2.689 if the drag force components are included in the equations of motion while the steady state eccentricity ratio is only dropped slightly from 0.250 to 0.247. When the instability threshold speeds are converted to a dimensional form, the instability threshold speed with consideration of the drag force is 11 288 rpm and the instability threshold speed neglecting the drag force becomes 11 094 rpm.

According to Chapter 4, since $\gamma_2 < 0$ and $\beta_2 > 0$ for both cases, unstable periodic solutions exist for the system running speeds, which are close to but less than the threshold speed $\bar{\omega}_{st}$. With the Hopf bifurcation parameters shown in Table 6.13 and the Hopf bifurcation formulae (Eqs. 4.2–4.4) given in Section 4.2, Figure 6.48 is plotted to compare the bifurcation profiles with and without consideration of the drag force effect.

Figure 6.48 compares the bifurcation profiles that depict the amplitudes of the unstable periodic solutions corresponding to a set of running speeds close to but less than the critical value $\bar{\omega}_{st}$ (instability threshold speed). The amplitude of the periodic solution at a specific running speed $\bar{\omega}$ is bounded by $\varepsilon_s \pm \sqrt{(\bar{\omega} - \bar{\omega}_{st})/\gamma_2}$. It shrinks to one point $\varepsilon_s(\bar{\omega}_{st})$ as the system running speed approaches the critical value, $\bar{\omega}_{st}$. The dimensionless instability threshold speed ($\bar{\omega}_{st} = 2.643$) without inclusion of the drag force components in the equations of motion is significantly lower than that ($\bar{\omega}_{st} = 2.689$) with inclusion of the drag force

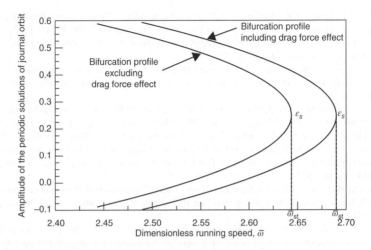

Figure 6.48 Bifurcation profiles. From Wang and Khonsari (2005) © SAGE Publications Ltd.

components. The difference between these two bifurcation profiles grows bigger as the system running speed $\bar{\omega}$ approaches the instability threshold speed $\bar{\omega}_{st}$.

The drag force effects on the stability of rotor-bearing system presented above are consistent with the qualitative explanation of "origin and development of the oil whirl" in a classical paper by Newkirk and Grobel (1934). Newkirk and Grobel concluded that a force component with a direction opposite to the direction of motion tended to stabilize the system. Equation 6.118 shows that the tangential component of the drag force (always negative since $(1-\varepsilon^3) > (1-\varepsilon^2)^{3/2}$ except at zero eccentricity ratio, it becomes zero when the shaft journal center coincides with the bearing bushing center) is always in the direction opposite to the direction of motion. Therefore, the inclusion of the drag force effects always makes the rotor-bearing system more stable.

References

Adams, M.L., Guo, J.S., 1996, "Simulations and Experiments of the Non-linear Hysteresis Loop for Rotor-Bearing Instability," IMechE Conference Transactions 1996–1996, Sixth International Conference on Vibration in Rotating Machinery, September 9-12, C500/001/96, Mechanical Engineering Publications Limited, London, pp. 309–319.

Akers, A., Michaelson, S., Cameron, A., 1971, "Stability Contours for a Finite Whirling Journal Bearing," ASME Journal of Lubrication Technology, **93**, pp. 177–190.

Bar-Yoseph, P., Blech, J.J., 1977, "Stability of a Flexible Rotor Supported by Circumferentially Fed Journal Bearings," Journal of Lubrication Technology, Transactions ASME, **99** (4), pp. 469–477.

Cameron, A., 1966, Principles of Lubrication, John Wiley & Sons, Inc., New York.

Dubois, G.B., Ocvirk, F.W., 1953, "Analytical Derivation and Experimental Evaluation of Short Bearing Approximation of Full Journal Bearings," U.S Government Printing Office, Ithaca, NY, USA, N.A.C.A. Technical Report 1157, pp. 1199–1230.

Dufrane, K.F., Kannel, J.W., McCloskey, T.H., 1983, "Wear of Steam Turbine Journal Bearings at Low Operating Speeds," *Journal of Lubrication Technology, Transactions ASME*, **105** (3), pp. 313–317.

Fillon, M., Bouyer, J., 2004, "Thermohydrodynamic Analysis of a Worn Plain Journal Bearing," *Tribology International*, **37** (2), pp. 129–136.

Guo, J.S., 1995, "Characteristics of the Nonlinear Hysteresis Loop for Rotor-Bearing Instability," Dissertation, Department of Mechanical and Aerospace Engineering, Case Western Reserve University, Cleveland, OH.

Guo, J.S., Adams, M.L., 1995, "Characteristics of the Nonlinear Hysteresis Loop for Rotor-Bearing Instability," DE-Vol.84-2, Proceedings of 1995 ASME Design Engineering Technical Conferences, 3 (Part B), Boston, MA, September 17-20, published by ASME, New York. pp. 1081–1091.

Hashimoto, H., Wada, S., Nojima, K., 1986a, "Performance Characteristic of Worn Journal Bearings in Both Laminar and Turbulent Regimes. Part I: Steady-State Characteristics," *ASLE Transactions*, **29** (4), pp. 565–571.

Hashimoto, H., Wada, S., Nojima, K., 1986b, "Performance Characteristic of Worn Journal Bearings in Both Laminar and Turbulent Regimes. Part II: Dynamic Characteristics," *ASLE Transactions*, **29** (4), pp. 572–577.

Hashimoto, H., Wada, S., Ito, J.I., 1987, "An Application of Short Bearing Theory to Dynamic Characteristic Problems of Turbulent Journal Bearings," *ASME Journal of Tribology*, **109**, pp. 307–314.

Hassard, B.D., Kazarinoff, N.D., Wan, Y.H., 1981, *Theory and Applications of Hopf Bifurcation*, London Mathematical Society Lecture Notes 41, Cambridge University Press, New York.

Hollis, P., Taylor, D.L., 1986, "Hopf Bifurcation to Limit Cycles in Fluid Film Bearings," *ASME Journal of Tribology*, **108**, pp. 184–189.

Horattas, G.A., Adams, M.L., Abdel Magied, M.F., and Loparo, K.A., 1997, "Experimental Investigation of Dynamic Nonlinearities in Rotating Machinery," Proceedings of the ASME Design Engineering Technical Conferences, September 14–17, Sacramento, CA, pp. 1–12.

Hori, Y., Kato, T., 1990, "Earthquake-Induced Instability of a Rotor Supported by Oil Film Bearings," *ASME Journal of Vibration and Acoustics*, **112**, pp. 160–165.

Khonsari, M.M., Booser, E.R., 2008, *Applied Tribology: Bearing Design and Lubrication*, 2nd edition, John Wiley & Sons, Ltd, Chichester, UK.

Kumar, A., Mishra, S.S, 1996a, "Stability of a Rigid Rotor in Turbulent Hydrodynamic Worn Journal Bearings," *Wear*, **193** (1), pp. 25–30.

Kumar, A., Mishra, S.S., 1996b,"Steady State Analysis of Non-circular Worn Journal Bearings in Nonlaminar Lubrication Regimes," *Tribology International*, **29** (6), pp. 493–498.

Lund, J.W., 1965, "The Stability of an Elastic Rotor in Journal Bearings with Flexible, Damped Supports," *ASME Journal of Applied Mechanics*, **87**, pp. 911–920.

Lundholm, G., 1971, *"The Axial Groove Journal Bearing Considering Cavitation and Dynamic Stability,"* Acta Polytechnica Scandinavica, Mechanical Engineering Series No. 58, Royal Swedish Academy of Engineering Sciences, Stockholm.

Maki, E.R., Ezzat, H.A., 1980, "Thermally Induced Whirl of a Rigid Rotor on Hydrodynamic Journal Bearing," *ASME Journal of Lubrication Technology*, **102**, pp. 8–14.

Mitchell, J.R., Holmes, R., Byrne, J., 1965, "Oil Whirl of a Rigid Rotor in 360 degree Journal Bearings: Further Characteristics," *Proceedings of the Institute of Mechanical Engineers*, **180** (25), pp. 593–610.

Newkirk, B.L., 1956, "Varieties of Shaft Disturbances Due to Fluid Films in Journal Bearings," *Transactions of the ASME*, **78**, pp. 985–988.

Newkirk, B.L., Grobel, L.P., 1934, Oil-Film Whirl—A Non-Whirling Bearing, *ASME Journal of Applied Mechanics*, **1**, pp. 607–615.

Newkirk, B.L., Lewis, J.F., 1956, "Oil Film Whirl—An Investigation of Disturbances Due to Oil Films in Journal Bearings," *Transactions of the ASME*, **78**, pp. 21–27.

Noah, S.T., 1995, "Significance of Considering Nonlinear Effects in Predicting the Dynamic Behavior of Rotating Machinery," *Journal of Vibration and Control*, **1**, pp. 431–458.

Pinkus, O., 1953, "Note on Oil Whip," *ASME Journal of Applied Mechanics*, **75**, pp. 450–451.

Pinkus, O., 1956, "Experimental Investigation of Resonant Whip," *Transactions of the ASME*, **78**, pp. 975–983.

Raimondi, A., Szeri, A.Z., 1984, "Journal and Thrust Bearings," *CRC Handbook of Lubrication: Vol. II Theory and Design*, Booser, E.R. ed., CRC Press, Boca Raton, FL, pp. 413–462.

Suzuki, K., Tanaka, M., 2002, "The Stability Characteristics of Two-lobe Journal Bearings with Surface Wear Dents: 2nd Report, Experimental Verification and Nonlinear Vibration Analysis," *Transactions of the Japan Society of Mechanical Engineers, Part C*, **68** (5), pp. 1447–1452.

Tanaka, M., Suzuki, K., 2002, "The Stability Characteristics of Two-Lobe Journal Bearings with Surface Wear Dents (1st Report, Theoretical Analysis)," *Transactions of the Japan Society of Mechanical Engineers, Part C*, **68** (5), pp. 1441–1446.

Vaidyanathan, K., Keith, T.G., 1991, "Performance Characteristics of Cavitated Noncircular Journal Bearings in the Turbulent Flow Regime," *Tribology Transactions*, **34** (1), pp. 35–44.

Wang, J.K., Khonsari, M.M., 2005, "Influence of Drag Force on the Dynamic Performance of Rotor-Bearing System," *Proceedings of the Institution of Mechanical Engineers, Part J, Journal of Engineering Tribology*, **219** (4), pp. 291–295.

Wang, J.K., Khonsari, M.M., 2006a, "Application of Hopf Bifurcation Theory to the Rotor-Bearing System with Turbulent Effects," *Tribology International*, **39** (7), pp. 701–714.

Wang, J.K., Khonsari, M.M., 2006b, "On the Hysteresis Phenomenon Associated with Instability of Rotor-Bearing Systems," *ASME Journal of Tribology*, **128**, pp. 188–196.

Wang, J.K., Khonsari, M.M., 2006c, "Prediction of the Stability Envelope of Rotor-Bearing System," *ASME Journal of Vibration and Acoustics*, **128**, pp. 197–202.

Wang, J.K., Khonsari, M.M., 2006d, "Influence of Inlet Oil Temperature on the Instability Threshold of Rotor-Bearing System," *ASME Journal of Tribology*, **128**, pp. 319–326.

Wang, J.K., Khonsari, M.M., 2006e, "Bifurcation Analysis of a Flexible Rotor Supported by Two Fluid Film Journal Bearings," *ASME Journal of Tribology*, **128**, pp. 594–603.

Wang, J.K., Khonsari, M.M., 2008a, "Effects of Oil Inlet Pressure and Inlet Position of Axially Grooved Infinitely Long Journal Bearings, Part I: Analytical Solutions and Static Performance," *Tribology International*, **41**, pp. 119–131.

Wang, J.K., Khonsari, M.M., 2008b, "Effects of Oil Inlet Pressure and Inlet Position of Axially Grooved Infinitely Long Journal Bearings, Part II: Nonlinear Instability Analysis," *Tribology International*, **41**, pp. 132–140.

Yu, L., Liu, H., 1996, *Rotor-Bearing System Dynamics*, Theory of Lubrication and Bearing Institute, Xi'an.

Zhang, Y., 1989, "Dynamic Properties of Flexible Journal Bearings of Infinite Width Considering Oil Supply Position and Pressure," *Wear*, **130** (1), pp. 53–68.

Appendix A

Derivation of the Dynamic Pressure for Long Journal Bearing

Integrating Equation 1.35 once with respect to θ, the following equation can be obtained (Wang and Khonsari, 2008):

$$\frac{\partial p}{\partial \theta} = 6\mu C^{-2} R^2 \left[\left(\omega - 2\dot{\phi} \right) \varepsilon \frac{\cos\theta}{\left(1 + \varepsilon\cos\theta\right)^3} + 2\dot{\varepsilon} \frac{\sin\theta}{\left(1 + \varepsilon\cos\theta\right)^3} + C_1 \frac{1}{\left(1 + \varepsilon\cos\theta\right)^3} \right]$$

(A.1)

where $C_1 \neq C_1(\theta)$.

Applying Equation 1.31 (i.e., $\partial p / \partial \theta = 0$ at $\theta = \theta_c$) to Equation (A.1) yields

$$C_1 = - \left(\omega - 2\dot{\phi} \right) \varepsilon\cos\theta_c - 2\dot{\varepsilon}\sin\theta_c$$

(A.2)

Thermohydrodynamic Instability in Fluid-Film Bearings, First Edition.
J. K. Wang and M. M. Khonsari.
© 2016 John Wiley & Sons, Ltd. Published 2016 by John Wiley & Sons, Ltd.

Integrating Equation (A.1) with C_1 defined by Equation (A.2), the following expression for pressure distribution is obtained:

$$p = 6\mu C^{-2} R^2 \left[(\omega - 2\dot{\phi})\varepsilon \int \frac{\cos\theta}{(1 + \varepsilon\cos\theta)^3} d\theta + 2\dot{\varepsilon} \int \frac{\sin\theta}{(1 + \varepsilon\cos\theta)^3} d\theta \right.$$

$$\left. - (\omega\varepsilon\cos\theta_c - 2\dot{\phi}\varepsilon\cos\theta_c + 2\dot{\varepsilon}\sin\theta_c) \int \frac{1}{(1 + \varepsilon\cos\theta)^3} d\theta + C_2 \right]$$

(A.3)

where $C_2 \neq C_2(\theta)$.

Solving the integrals (see Appendix B) and rearranging the results, Equation (A.3) is rewritten as follows:

$$p = \frac{6\mu R^2}{C^2 (1 - \varepsilon^2)^{1.5}} \left[(\omega - 2\dot{\phi})(\alpha - \varepsilon\sin\alpha) + \frac{\dot{\varepsilon}(1 - \varepsilon\cos\alpha)^2}{\varepsilon(1 - \varepsilon^2)^{0.5}} \right.$$

$$\left. - \frac{(\omega - 2\dot{\phi})(1 + \varepsilon\cos\theta_c) + 2\dot{\varepsilon}\sin\theta_c}{1 - \varepsilon^2} \left(\alpha + \frac{\varepsilon^2}{2}\alpha - 2\varepsilon\sin\alpha + \frac{\varepsilon^2\sin2\alpha}{4} \right) + C_2 \right]$$

(A.4)

Using Sommerfeld's substitutions $d\theta = \left((1 - \varepsilon^2)^{0.5} / (1 - \varepsilon\cos\alpha) \right) d\alpha$ and $1 + \varepsilon\cos\theta = (1 - \varepsilon^2)/(1 - \varepsilon\cos\alpha)$, Equations A.5–A.7 are derived as follows (Wang and Khonsari, 2008):

$$\cos\theta_s = \frac{\cos\alpha_s - \varepsilon}{1 - \varepsilon\cos\alpha_s}$$

$$\sin\theta_s = \frac{(1 - \varepsilon^2)^{0.5}\sin\alpha_s}{1 - \varepsilon\cos\alpha_s}$$

(A.5)

$$\cos\theta_c = \frac{\cos\alpha_c - \varepsilon}{1 - \varepsilon\cos\alpha_c}$$

$$\sin\theta_c = \frac{(1 - \varepsilon^2)^{0.5}\sin\alpha_c}{1 - \varepsilon\cos\alpha_c}$$

(A.6)

$$\cos\alpha_i = \frac{\varepsilon + \cos\theta_i}{1 + \varepsilon\cos\theta_i} = \frac{\varepsilon + \cos(\Theta_i - \phi)}{1 + \varepsilon\cos(\Theta_i - \phi)}$$

$$\sin\alpha_i = \frac{(1 - \varepsilon^2)^{0.5}\sin\theta_i}{1 + \varepsilon\cos\theta_i} = \frac{(1 - \varepsilon^2)^{0.5}\sin(\Theta_i - \phi)}{1 + \varepsilon\cos(\Theta_i - \phi)}$$

(A.7)

where α_s corresponds to θ_s that defines the oil pressure starting position and α_c corresponds to θ_c that represents the starting position of the cavitation and α_i corresponds to θ_i that specifies the oil inlet position.

Substituting Equation A.6 into Equation A.4 yields

$$
p = \frac{6\mu R^2}{C^2(1-\varepsilon^2)^2}\left[(\omega - 2\dot{\phi})(1-\varepsilon^2)^{0.5}(\alpha - \varepsilon\sin\alpha) + \frac{\dot{\varepsilon}(1-\varepsilon\cos\alpha)^2}{\varepsilon}\right.
$$
$$
\left. - \frac{(\omega - 2\dot{\phi})(1-\varepsilon^2)^{0.5} + 2\dot{\varepsilon}\sin\alpha_c}{(1-\varepsilon\cos\alpha_c)}\left(\alpha + \frac{\varepsilon^2}{2}\alpha - 2\varepsilon\sin\alpha + \frac{\varepsilon^2\sin2\alpha}{4}\right) + C_2\right]
$$

(A.8)

where $C_2 \neq C_2(\alpha)$ since $C_2 \neq C_2(\theta)$. Equation A.8 is an analytical expression for the hydrodynamic pressure p.

C_2 in Equation A.8 is determined through applying the boundary condition at the oil inlet position—Equation 1.33—into Equation A.8.

$$
C_2 = \frac{C^2(1-\varepsilon^2)^2}{6\mu R^2}p_i - (\omega - 2\dot{\phi})(1-\varepsilon^2)^{0.5}(\alpha_i - \varepsilon\sin\alpha_i) - \frac{\dot{\varepsilon}(1-\varepsilon\cos\alpha_i)^2}{\varepsilon}
$$
$$
+ \frac{(\omega - 2\dot{\phi})(1-\varepsilon^2)^{0.5} + 2\dot{\varepsilon}\sin\alpha_c}{(1-\varepsilon\cos\alpha_c)}\left(\alpha_i + \frac{\varepsilon^2}{2}\alpha_i - 2\varepsilon\sin\alpha_i + \frac{\varepsilon^2\sin2\alpha_i}{4}\right)
$$

(A.9)

Substituting Equation A.9 into Equation A.8, the analytical expression for the hydrodynamic pressure p becomes (Wang and Khonsari, 2008)

$$
p = p_i + \frac{3\mu R^2}{C^2(1-\varepsilon^2)^2}\left\{\frac{(\omega - 2\dot{\phi})\varepsilon(1-\varepsilon^2)^{0.5}}{(1-\varepsilon\cos\alpha_c)}[2(\sin\alpha - \sin\alpha_i)(1 + \varepsilon\cos\alpha_c)\right.
$$
$$
+ (2\cos\alpha_c + \varepsilon)(\alpha_i - \alpha) - \varepsilon(\sin\alpha\cos\alpha - \sin\alpha_i\cos\alpha_i)]
$$
$$
+ \dot{\varepsilon}\frac{2\sin\alpha_c}{(1-\varepsilon\cos\alpha_c)}\left[\frac{(2 - \varepsilon\cos\alpha - \varepsilon\cos\alpha_i)(\cos\alpha_i - \cos\alpha)(1 - \varepsilon\cos\alpha_c)}{\sin\alpha_c}\right.
$$
$$
\left.\left. - (2 + \varepsilon^2)(\alpha - \alpha_i) + 4\varepsilon(\sin\alpha - \sin\alpha_i) - \varepsilon^2(\sin\alpha\cos\alpha - \sin\alpha_i\cos\alpha_i)\right]\right\}
$$

(A.10)

Reference

Wang, J.K., Khonsari, M.M., 2008, "Effects of Oil Inlet Pressure and Inlet Position of Axially Grooved Infinitely Long Journal Bearings, Part I: Analytical Solutions and Static Performance", *Tribology International*, **41**, February, pp. 119–131.

Appendix B

Integrals Used in Section 1.3

Assume integrals (Wang and Khonsari, 2008)

$$J_l^{mn} = \int \frac{\sin^m\theta\cos^n\theta}{(1+\varepsilon\cos\theta)^l}d\theta \tag{B.1}$$

Although integrals J_l^{mn} can be evaluated using the formulas provided by Booker (1965), here one substitution introduced by Dubois and Ocvirk (1953) and also Sommerfeld's substitution are used to determine the integrals J_l^{mn}.

The following substitution (Dubois and Ocvirk, 1953) is used frequently in the integrations:

$$\frac{\cos\theta}{(1+\varepsilon\cos\theta)} = \frac{1}{\varepsilon}\left[1 - \frac{1}{(1+\varepsilon\cos\theta)}\right] \tag{B.2}$$

Using Sommerfeld's substitutions $d\theta = \left((1-\varepsilon^2)^{0.5}/(1-\varepsilon\cos\alpha)\right)d\alpha$ and $1+\varepsilon\cos\theta = (1-\varepsilon^2)/(1-\varepsilon\cos\alpha)$, one has the following set of infinite integrations (Wang and Khonsari, 2008):

$$J_3^{00} = \frac{1}{(1-\varepsilon^2)^{2.5}}\left(\alpha + \frac{\varepsilon^2}{2}\alpha - 2\varepsilon\sin\alpha + \frac{\varepsilon^2\sin2\alpha}{4}\right) \tag{B.3}$$

Thermohydrodynamic Instability in Fluid-Film Bearings, First Edition.
J. K. Wang and M. M. Khonsari.
© 2016 John Wiley & Sons, Ltd. Published 2016 by John Wiley & Sons, Ltd.

$$J_3^{10} = \frac{(1 - \varepsilon\cos\alpha)^2}{2\varepsilon(1 - \varepsilon^2)^2} \tag{B.4}$$

$$J_3^{01} = \frac{-3\varepsilon\alpha + 2(1 + \varepsilon^2)\sin\alpha - \varepsilon\sin\alpha\cos\alpha}{2(1 - \varepsilon^2)^{2.5}} \tag{B.5}$$

$$J_3^{11} = \frac{(1 - 2\varepsilon^2 + \varepsilon\cos\alpha)(1 - \varepsilon\cos\alpha)}{2\varepsilon^2(1 - \varepsilon^2)^2} \tag{B.6}$$

$$J_3^{20} = \frac{\alpha - \sin\alpha\cos\alpha}{2(1 - \varepsilon^2)^{1.5}} \tag{B.7}$$

$$J_3^{02} = \frac{(1 + 2\varepsilon^2)\alpha - 4\varepsilon\sin\alpha + \sin\alpha\cos\alpha}{2(1 - \varepsilon^2)^{2.5}} \tag{B.8}$$

References

Booker, J.F., 1965, "A Table of the Journal-Bearing Integral," *ASME Journal of Basic Engineering*, **87**, pp. 533–535.

Dubois, G.B., Ocvirk, F.W., 1953, Analytical derivation and experimental evaluation of short bearing approximation of full journal bearings. NACA technical report 1157. U.S. Government Printing Office. Ithaca, NY, USA.

Wang, J.K., Khonsari, M.M., 2008, "Effects of Oil Inlet Pressure and Inlet Position of Axially Grooved Infinitely Long Journal Bearings, Part I: Analytical Solutions and Static Performance," *Tribology International*, **41**, February, pp. 119–131.

Appendix C

Curve-fitting Functions for Long Journal Bearings

Based on the results presented in Figures 1.7, 1.8, 1.9, 1.10, 1.11, 1.12, 1.13, and 1.14, the following sets of curve-fitting functions are obtained and given in Equations C.1–C.20. Under different oil supply conditions, Equations C.1–C.20 reveal how θ_c (circumferential location where the oil film ruptures) and θ_s (oil pressure starting position) and ϕ (attitude angle) change with changing the steady-state eccentricity ratio ε and how the steady-state eccentricity ratio ε changes with changing the Sommerfeld number (S) (Wang and Khonsari, 2008).

Case I: Oil Supply Condition ($\Theta_i = 0$, $\bar{p}_i = 0$)

$$\theta_c = -141.27\varepsilon^4 + 182.27\varepsilon^3 - 140.49\varepsilon^2 - 64.670\varepsilon + 359.11 \tag{C.1}$$

$$\theta_s = 0 \tag{C.2}$$

$$\phi = -80.378\varepsilon^4 + 54.495\varepsilon^3 - 61.798\varepsilon^2 + 8.8284\varepsilon + 89.597 \tag{C.3}$$

$$\varepsilon = \begin{cases} 573330S^4 - 39933S^3 + 630.99S^2 - 12.268S + 1.0068 & S \leq 0.04 \\ 0.016816S^{-1.0048} & S > 0.04 \end{cases} \tag{C.4}$$

Thermohydrodynamic Instability in Fluid-Film Bearings, First Edition.
J. K. Wang and M. M. Khonsari.
© 2016 John Wiley & Sons, Ltd. Published 2016 by John Wiley & Sons, Ltd.

Case II: Oil Supply Condition ($\Theta_i = 0$, $\bar{p}_i = 0.5$)

If $\varepsilon > 0.4$, cavitation exists.

$$\theta_c = -1263.1\varepsilon^4 + 3201.7\varepsilon^3 - 3106.4\varepsilon^2 + 1183.3\varepsilon + 176.46 \qquad (C.5)$$

$$\theta_s = -1214.0\varepsilon^4 + 3662.3\varepsilon^3 - 4159.7\varepsilon^2 + 2096.4\varepsilon - 44.228 \qquad (C.6)$$

No cavitation exists when $\varepsilon \leq 0.4$.

$$\phi = \begin{cases} -815.63\varepsilon^4 + 2031.6\varepsilon^3 - 2012.1\varepsilon^2 + 838.24\varepsilon - 33.149 & \varepsilon > 0.4 \\ 90° & \varepsilon \leq 0.4 \end{cases} \qquad (C.7)$$

$$\varepsilon = \begin{cases} 682040S^4 - 45006S^3 + 642.43S^2 - 11.754S + 1.0050 & S \leq 0.04 \\ 0.016916S^{-0.99816} & S > 0.04 \end{cases} \qquad (C.8)$$

Case III: Oil Supply Condition ($\Theta_i = 0$, $\bar{p}_i = 1.0$)

If $\varepsilon > 0.49$, cavitation exists.

$$\theta_c = -2368.3\varepsilon^4 + 6522.9\varepsilon^3 - 6811.8\varepsilon^2 + 2991.1\varepsilon - 142.76 \qquad (C.9)$$

$$\theta_s = -2545.6\varepsilon^4 + 7956.5\varepsilon^3 - 9400.3\varepsilon^2 + 4945.9\varepsilon - 631.61 \qquad (C.10)$$

No cavitation exists when $\varepsilon \leq 0.49$.

$$\phi = \begin{cases} -1609.5\varepsilon^4 + 4407.0\varepsilon^3 - 4662.1\varepsilon^2 + 2136.6\varepsilon - 263.82 & \varepsilon > 0.49 \\ 90° & \varepsilon \leq 0.49 \end{cases} \qquad (C.11)$$

$$\varepsilon = \begin{cases} 708090S^4 - 43282S^3 + 493.53S^2 - 9.9516S + 1.0010 & S \leq 0.04 \\ 0.016917S^{-0.99814} & S > 0.04 \end{cases} \qquad (C.12)$$

Case IV: Oil Supply Condition ($\Theta_i = 45°$, $\bar{p}_i = 0$)

$$\theta_c = -248.84\varepsilon^4 + 393.49\varepsilon^3 - 205.68\varepsilon^2 - 91.309\varepsilon + 347.20 \qquad (C.13)$$

$$\theta_s = 45° \qquad (C.14)$$

$$\phi = -168.09\varepsilon^4 + 259.14\varepsilon^3 - 160.81\varepsilon^2 - 0.95396\varepsilon + 81.696 \tag{C.15}$$

$$\varepsilon = \begin{cases} 56139S^4 - 4143.4S^3 + 3.2588S^2 - 8.1780S + 1.0008 & S \le 0.04 \\ 0.017752S^{-1.0620} & S > 0.04 \end{cases} \tag{C.16}$$

Case V: Oil Supply Condition ($\Theta_i = 90°$, $\bar{p}_i = 0$)

$$\theta_c = -220.28\varepsilon^4 + 334.49\varepsilon^3 - 165.64\varepsilon^2 - 60.425\varepsilon + 306.37 \tag{C.17}$$

$$\theta_s = 90° \tag{C.18}$$

$$\phi = -157.25\varepsilon^4 + 242.52\varepsilon^3 - 135.91\varepsilon^2 + 2.3925\varepsilon + 58.834 \tag{C.19}$$

$$\varepsilon = \begin{cases} -2436.6S^4 + 679.69S^3 - 32.682S^2 - 7.4427S + 0.99923 & S \le 0.1 \\ 0.035514S^{-1.0150} & S > 0.1 \end{cases} \tag{C.20}$$

Reference

Wang, J.K., Khonsari, M.M., 2008, "Effects of Oil Inlet Pressure and Inlet Position of Axially Grooved Infinitely Long Journal Bearings, Part I: Analytical Solutions and Static Performance", *Tribology International*, **41**, February, pp. 119–131.

Appendix D

Jacobian Matrix of the Equations of Motion

The Jacobian matrix of the equations of motion (Eq. 6.34) is defined as

$$\frac{\partial \mathbf{f}}{\partial \mathbf{x}}(\mathbf{x}_s, \omega) = \begin{bmatrix} \dfrac{\partial f_1}{\partial x_1} & \dfrac{\partial f_1}{\partial x_2} & \dfrac{\partial f_1}{\partial x_3} & \dfrac{\partial f_1}{\partial x_4} & \dfrac{\partial f_1}{\partial x_5} & \dfrac{\partial f_1}{\partial x_6} \\[2ex] \dfrac{\partial f_2}{\partial x_1} & \dfrac{\partial f_2}{\partial x_2} & \dfrac{\partial f_2}{\partial x_3} & \dfrac{\partial f_2}{\partial x_4} & \dfrac{\partial f_2}{\partial x_5} & \dfrac{\partial f_2}{\partial x_6} \\[2ex] \dfrac{\partial f_3}{\partial x_1} & \dfrac{\partial f_3}{\partial x_2} & \dfrac{\partial f_3}{\partial x_3} & \dfrac{\partial f_3}{\partial x_4} & \dfrac{\partial f_3}{\partial x_5} & \dfrac{\partial f_3}{\partial x_6} \\[2ex] \dfrac{\partial f_4}{\partial x_1} & \dfrac{\partial f_4}{\partial x_2} & \dfrac{\partial f_4}{\partial x_3} & \dfrac{\partial f_4}{\partial x_4} & \dfrac{\partial f_4}{\partial x_5} & \dfrac{\partial f_4}{\partial x_6} \\[2ex] \dfrac{\partial f_5}{\partial x_1} & \dfrac{\partial f_5}{\partial x_2} & \dfrac{\partial f_5}{\partial x_3} & \dfrac{\partial f_5}{\partial x_4} & \dfrac{\partial f_5}{\partial x_5} & \dfrac{\partial f_5}{\partial x_6} \\[2ex] \dfrac{\partial f_6}{\partial x_1} & \dfrac{\partial f_6}{\partial x_2} & \dfrac{\partial f_6}{\partial x_3} & \dfrac{\partial f_6}{\partial x_4} & \dfrac{\partial f_6}{\partial x_5} & \dfrac{\partial f_6}{\partial x_6} \end{bmatrix} \tag{D.1}$$

Thermohydrodynamic Instability in Fluid-Film Bearings, First Edition.
J. K. Wang and M. M. Khonsari.
© 2016 John Wiley & Sons, Ltd. Published 2016 by John Wiley & Sons, Ltd.

Substituting the equations of motion (Eq. 6.34) into the Jacobian matrix, we obtain the following (Wang and Khonsari, 2006):

$$\frac{\partial \mathbf{f}}{\partial \mathbf{x}}(\mathbf{x}_s, \omega) = \begin{bmatrix} J_{11} & J_{12} & J_{13} & 0 & J_{15} & 0 \\ J_{21} & J_{22} & J_{23} & 0 & J_{25} & 0 \\ 0 & 0 & 0 & 1 & 0 & 0 \\ \dfrac{\bar{K}}{\bar{\omega}^2}\sin x_2 & \dfrac{\bar{K}}{\bar{\omega}^2}x_1\cos x_2 & -\dfrac{\bar{K}}{\bar{\omega}^2} & 0 & 0 & 0 \\ 0 & 0 & 0 & 0 & 0 & 1 \\ -\dfrac{\bar{K}}{\bar{\omega}^2}\cos x_2 & \dfrac{\bar{K}}{\bar{\omega}^2}x_1\sin x_2 & 0 & 0 & -\dfrac{\bar{K}}{\bar{\omega}^2} & 0 \end{bmatrix} \qquad (D.2)$$

where

$$J_{11} = -\frac{4(\pi^2-8)\bar{K}x_1\left[\pi\left(1-x_1^2\right)^{0.5}(x_3\sin x_2 - x_5\cos x_2 - x_1) + 4x_1(x_3\cos x_2 + x_5\sin x_2)\right]}{\bar{\omega}\Gamma\left(1-x_1^2\right)^{-2}\left[\pi^2 + 2(\pi^2-8)x_1^2\right]^2}$$
$$- \frac{\bar{K}\left[5\pi x_1(x_3\sin x_2 - x_5\cos x_2 - x_1) + \pi\left(1-x_1^2\right)\right]}{\bar{\omega}\Gamma\left(1-x_1^2\right)^{-1.5}\left[\pi^2 + 2(\pi^2-8)x_1^2\right]} + \frac{4\bar{K}\left(1-5x_1^2\right)(x_3\cos x_2 + x_5\sin x_2)}{\bar{\omega}\Gamma\left(1-x_1^2\right)^{-1}\left[\pi^2 + 2(\pi^2-8)x_1^2\right]}$$

$$J_{12} = \frac{\bar{K}\left[\pi\left(1-x_1^2\right)^{2.5}(x_3\cos x_2 + x_5\sin x_2) + 4x_1\left(1-x_1^2\right)^2(x_5\cos x_2 - x_3\sin x_2)\right]}{\bar{\omega}\Gamma\left[\pi^2 + 2(\pi^2-8)x_1^2\right]}$$

$$J_{13} = \frac{\bar{K}\left[\pi\left(1-x_1^2\right)^{2.5}\sin x_2 + 4x_1\left(1-x_1^2\right)^2\cos x_2\right]}{\bar{\omega}\Gamma\left[\pi^2 + 2(\pi^2-8)x_1^2\right]}$$

$$J_{15} = \frac{\bar{K}\left[-\pi\left(1-x_1^2\right)^{2.5}\cos x_2 + 4x_1\left(1-x_1^2\right)^2\sin x_2\right]}{\bar{\omega}\Gamma\left[\pi^2 + 2(\pi^2-8)x_1^2\right]}$$

$$J_{21} = -\frac{4(\pi^2-8)\bar{K}\left[4x_1\left(1-x_1^2\right)^{0.5}(x_3\sin x_2 - x_5\cos x_2 - x_1) + \pi\left(1+2x_1^2\right)(x_3\cos x_2 + x_5\sin x_2)\right]}{\bar{\omega}\Gamma\left(1-x_1^2\right)^{-1.5}\left[\pi^2 + 2(\pi^2-8)x_1^2\right]^2}$$
$$- \frac{4\bar{K}\left[4x_1(x_3\sin x_2 - x_5\cos x_2) - 5x_1^2 + 1\right]}{\bar{\omega}\Gamma\left(1-x_1^2\right)^{-1}\left[\pi^2 + 2(\pi^2-8)x_1^2\right]} - \frac{\pi\bar{K}\left(1+8x_1^4\right)(x_3\cos x_2 + x_5\sin x_2)}{\bar{\omega}\Gamma x_1^2\left(1-x_1^2\right)^{-0.5}\left[\pi^2 + 2(\pi^2-8)x_1^2\right]}$$

$$J_{22} = \frac{\bar{K}\left[4\left(1-x_1^2\right)^2(x_3\cos x_2 + x_5\sin x_2) + \pi\left(1-x_1^2\right)^{1.5}((1/x_1) + 2x_1)(x_5\cos x_2 - x_3\sin x_2)\right]}{\bar{\omega}\Gamma\left[\pi^2 + 2(\pi^2-8)x_1^2\right]}$$

$$J_{23} = \frac{\bar{K}\left[4\left(1-x_1^2\right)^2\sin x_2 + \pi\left(1-x_1^2\right)^{1.5}((1/x_1) + 2x_1)\cos x_2\right]}{\bar{\omega}\Gamma\left[\pi^2 + 2(\pi^2-8)x_1^2\right]}$$

$$J_{25} = \frac{\bar{K}\left[-4\left(1-x_1^2\right)^2\cos x_2 + \pi\left(1-x_1^2\right)^{1.5}\left(\left(1/x_1\right)+2x_1\right)\sin x_2\right]}{\bar{\omega}\Gamma\left[\pi^2 + 2(\pi^2-8)x_1^2\right]}$$

Reference

Wang, J.K., Khonsari, M.M., 2006, "Bifurcation Analysis of a Flexible Rotor Supported by Two Fluid Film Journal Bearings," *ASME Journal of Tribology*, **128**, July, pp. 594–603.

$$\frac{1}{R} = 2(1 - r^2) \cos \ldots + r^2 - 4r^{2n} (r, r, \ldots) + \ldots}{c^2 (1 + r^2 - 2 \cos \ldots)}$$

References

1. Montelius, M. S. and C... "Interference in Analysis and Circular Rate Suppressed by TV Plant Experimental Studies", ASME, approved in Transactions TN 1986, pp. 46–67.

Appendix E

Matlab Code to Evaluate Rotor Shaft Unbalance Effects

E1 Main Code

```
clc;
clear all;
format long;
%mue=0.006 %unbalance input, unit: ounce.inch
%mue=mue*28.35e-3*25.4e-3;
mue=4/2;%unbalance input, unit: g.mm
mue=mue*1e-3*1e-3;
%mue=0
R=1/2*25.4/1000; %Diameter of the ends of the shaft
r=0.0*25.4/1000; %Modified to be used to calculate "I"
C=0.004*25.4/2/1000; % radial clearance
B=1/2*25.4/1000; % length of the bearing
m=12*0.4536; %m:the weight of whole rotor as a reference
only to check whether we get the correct mass calculation.
m=m/2;
g=9.81;
Wt=m*g;
L_D=B/(2*R);
% [kr,KR,FC,WK]=kshaft(R,r,C,B,m);
[FC,WK]=kshaft(R,r,C,B,m);
```

Thermohydrodynamic Instability in Fluid-Film Bearings, First Edition.
J. K. Wang and M. M. Khonsari.
© 2016 John Wiley & Sons, Ltd. Published 2016 by John Wiley & Sons, Ltd.

```
%Given the inlet oil temperature (degree C)
ti=40;
% Given running speed w (rpm).
%w=9000;% Note:de=0.01,the critical runup speed for orbit
from stable to unstable is just one point, not a range.
No limit cycle exists(Test must be run long enough.)
%w=;% Note:de=0.90,the critical rundown speed for orbit
from unstable to stable is just one point, not a range.
No limit cycle exists(Test must be run long enough.)
%w=%threshold speed de=0.001
w=8000

%Given perturbation
%de=0.058; % for runup threshold speed (in form of a
eccentricity perturbation)
de=0.001;
% dimensionless time span(dimensionless time=w(rad/s)*
dimensional time(sec))
tspan=[0,260];

N=w/60;
w=w*2*pi/60; %rad/s

% To get the viscosity at different temperature
v40=31.3;
v100=5.25;
D156=877;
temp=[0:1200];
B1=log10(log10(v40+0.7)./log10(v100+0.7))./log10
((273.15+100)./(273.15+40));
A=log10(log10(v40+0.7))+B1*log10(273.15+40);
for i=1:1201;
   v(i)=10^(10^(A-B1*log10(273.15+temp(i))))-0.7;
   D(i)=D156*(1-0.00063*(temp(i)-15.6));
   u(i)=v(i)*10^(-6)*D(i);
end;
T=temp*9/5+32;
tu(1:150,1)=temp(1:150)';
%datatu = fopen('tu.txt','w');
for i=1:91;
   ishaft=int16(28.90818+0.43985*i+0.0028*i^2);
   beta(i)=log(u(ishaft)/u(i))./(temp(i)-temp(ishaft));
end;
```

```
%for i=21:91
%   tu(i,2)=u(i);
%   %fprintf(datatu,'%6.0fC %7.4fPa.s %7.4f\n',tu(i,:),
beta(i));
%   %fprintf('%6.0fC %7.4fPa.s %7.4f\n',tu(i,:),beta(i));
%end;
%fclose(datatu);
grid;
%***************

%Calculate the viscosity of the effective oil film
temperature ( Consider the calculated shaft temperature as
the effective oil film temperature)
ti0=int16(ti);
ti1=int16(ti+1);
U=u(ti0)+(u(ti1)-u(ti0))*(mod(ti,1));
tshaft=ti;
beta=beta(ti0)+(beta(ti1)-beta(ti0))*(mod(ti,1))
alpha=0.76*10^(-7);
kc=0.13;
aa=0.41995;
bb=0.04391;
cc=0.12772;
k1=alpha/kc*U*beta*(R/C)^2*w;
k2=sqrt(U/kc*beta)*R*w;
dtshaft=(aa+bb/k1+cc/(k2^1.5))^(-1);
tshaft=dtshaft/beta+ti;
teffective=tshaft;
tt=int16(teffective);
tt1=int16(teffective+1);
mu=u(tt)+(u(tt1)-u(tt))*(mod(teffective,1))

% Using the given speed, calculate the Eccentricity of the
steady state equilibrium position
S=mu*w*B*R/(pi*(m*g))*(R/C)^2;

%S=0.7;% suppose mu*w=const and they can make S constant
at 0.7
%S=1;% suppose mu*w=const and they can make S constant at 1
%m=mu*w*B*R/(pi*g)*(R/C)^2/S;

wd=w*sqrt(C/g); % dimensionless running speed
lambda=((WK*sqrt(C/g))/wd)^2;
```

```
gamma=mu*R*B^3/(2*m*C^2.5*g^0.5);
delta=mue/(m*C)

i=1;
es(i)=0.9;% initial guess
des=0.1;% initial guess
while abs(des)>1e-10
   % Using Newton Iteration method to calculate the
eccentricity using the Sommerfeld No. S
%    f=(pi^2*(1-es(i)^2)+16*es(i)^2)*(pi*es(i)*S)^2-(1-es
(i)^2)^4*(2*R/B)^4;
%    fp=2*(pi^2*(1-es(i)^2)+16*es(i)^2)*(pi*S)^2*es(i)
+(pi*es(i)*S)^2*(32*es(i)-2*pi^2*es(i))+8*es(i)*(1-es
(i)^2)^3*(2*R/B)^4;
   % Alternative derivation
   f=es(i)*(pi^2+2*(pi^2-8)*es(i)^2)*sqrt(16*es(i)^2+
pi^2*(1-es(i)^2))+2/(2*pi*S*L_D^2)*(1-es(i)^2)^2*(16*es
(i)^2-pi^2*(1+2*es(i)^2));
fp=(pi^2+2*(pi^2-8)*es(i)^2)*sqrt(16*es(i)^2+pi^2*(1-es
(i)^2))+es(i)*(4*(pi^2-8)*es(i)*sqrt(16*es(i)^2+pi^2*(1-
es(i)^2))+0.5*(32*es(i)-2*pi^2*es(i))*(pi^2+2*(pi^2-8)*es
(i)^2)*(16*es(i)^2+pi^2*(1-es(i)^2))^-0.5);
   fp=fp-8*es(i)/(2*pi*S*L_D^2)*(1-es(i)^2)*(16*es(i)^2-
pi^2*(1+2*es(i)^2))+2/(2*pi*S*L_D^2)*(1-es(i)^2)^2*(32*es
(i)-4*pi^2*es(i));
   es(i+1)=es(i)-f/fp;
   des=es(i+1)-es(i);
   i=i+1;
end;
[dum, nes]=size(es);
if es(nes)>1
   'The input is out of range'
   break;
else
end
es0=es(nes)% Steady state equilibrium position
phis0=atan(pi/4*sqrt(1-es0^2)/es0);
xx0=es0*sin(phis0);
yy0=-es0*cos(phis0);

% start position
x10=es0+de
x20=atan(pi/4*sqrt(1-x10^2)/x10)
```

```
%x30=x10+FC*cos(x20/2) % initial condition for center
weight
x30=x10*sin(x20) % initial condition for center weight
(right below the perturbed position of the journal)
%x30=sqrt(es0^2+FC^2-2*es0*FC*cos(pi-phis0)) % initial
condition for center weight (right below the steady state
position of the journal)
x40=0.0
%x50=x20-x20/2
x50=-x10*cos(x20)-FC% initial condition for center weight
(right below the perturbed position of the journal)
%x50=phis0-acos((x30^2+es0^2-FC^2)/(2*x30*es0))%
initial condition for center weight (right below the steady
state position of the journal)
x60=0.0

% given start position
%x10=0.01
%x20=pi/(180/30); % range[0,360]
%x30=x10*sin(x20)
%x40=0
%x50=-x10*cos(x20)-FC
%x60=0

Bp=S/w*sqrt(g/C)
m_depsilon=64*Wt*m*C/(mu*2*R*B)^2*(C/B)^4

I=[x10,x20,x30,x40,x50,x60];

options=odeset('RelTol', 1e-10, 'AbsTol', 1e-10);
[t,v]=ode45(@whirl_fullflexiblewithunbalance,tspan,I,
options,wd,lambda,gamma,delta);
%[t,v]=ode45(@whirl_fullflexiblewithunbalance,tspan,I,
[],wd,lambda,gamma,delta);

x1=double(v(:,1));
x2=double(v(:,2));
xx(:,1)=x1(:,1).*sin(x2(:,1));
yy(:,1)=-x1(:,1).*cos(x2(:,1));
x3=double(v(:,3));
x5=double(v(:,5));
xxd(:,1)=x3(:,1);
yyd(:,1)=x5(:,1);
```

```
% Clearance circle
unitc1=[0:(2*pi/60):2*pi]';
[nunitc1,dum]=size(unitc1)
unitc2=ones(nunitc1,1);
unitx=cos(unitc1);
unity=sin(unitc1);

%the orbit of the journal
polar(unitc1,unitc2) % Unit circle
hold on;
polar(x2+3*pi/2,x1)
polar(x2(1,1)+3*pi/2,x1(1,1),'o')
title('Orbit of the journal')
hold off;

figure;
plot(xx,yy)
hold on;
text(xx(1,1), yy(1,1), 'start')
plot(xx(1,1),yy(1,1),'ko')
%plot(xx0,yy0,'k.')
%text(xx0, yy0, 'steady state position')
xlabel('Horizontal direction coordinate of the journal
x/C')
ylabel('Vertical direction coordinate of the journal y/C')
title('Orbit of the journal')
hold off;
grid;

figure;
plot(unitx,unity)
hold on;
plot(xx,yy)
plot(xxd,yyd)
%text(xx(1,1), yy(1,1), 'start')
%plot(xx(1,1),yy(1,1),'ko')
xlabel('Horizontal direction coordinate of the journal
x/C')
ylabel('Vertical direction coordinate of the journal y/C')
title('Orbit of the journal center and central disk center')
hold off;
grid;
```

```
figure;
plot(xxd,yyd)
hold on;
text(xxd(1,1), yyd(1,1), 'start')
plot(xxd(1,1),yyd(1,1),'ko')
xlabel('Horizontal direction coordinate of the disk x/C')
ylabel('Vertical direction coordinate of the disk y/C')
title('Orbit of the disk')
hold off;
grid;


[nt,dum]=size(t);
t_total=t(nt,1)


Undamped_1st_Natural_Frequency=WK/(2*pi)


figure
subplot(2,1,1),plot(t/w,xx);
xlabel('Time (sec)')
ylabel('Vertical direction coordinate of the journal
center x/C')
grid;
subplot(2,1,2),plot(t/w,yy);
xlabel('Time (sec)')
ylabel('Vertical direction coordinate of the journal
center y/C')
grid;
```

E2   Functions

E2.1   Function whirl_fullflexiblewithunbalance.m

```
function yprime=whirl_fullflexible(t,I,wd,lambda,
gamma,delta);
x1=I(1);
x2=I(2);
x3=I(3);
x4=I(4);
x5=I(5);
x6=I(6);
```

```
%Equation 1
X1=wd*lambda/gamma*(pi*(1-x1^2)^2.5*(x3*sin(x2)-x5*cos
(x2)-x1)+4*x1*(1-x1^2)^2*(x3*cos(x2)+x5*sin(x2)))/(pi^2
+2*(pi^2-8)*x1^2);
%Equation 2
X2=0.5+wd*lambda/gamma*(4*(1-x1^2)^2*(x3*sin(x2)-x5*cos
(x2)-x1)+pi*(1-x1^2)^1.5*(1/x1+2*x1)*(x3*cos(x2)+x5*sin
(x2)))/(pi^2+2*(pi^2-8)*x1^2);
%Equation 3
X3=x4;
%Equation 4
%for different initial phase angle, wt starts from x axis
%X4=-lambda*x3+lambda*x1*sin(x2)+delta*cos(t);
%X4=-lambda*x3+lambda*x1*sin(x2)+delta*cos(t-pi/2);

%for different initial phase angle, wt starts from -y axis
X4=-lambda*x3+lambda*x1*sin(x2)+delta*sin(t);
%Equation 5
X5=x6;
%Equation 6
%for different initial phase angle, wt starts from x axis
%X6=-lambda*x5-lambda*x1*cos(x2)-1/wd^2+delta*sin(t);
%X6=-lambda*x5-lambda*x1*cos(x2)-1/wd^2+delta*sin(t-
pi/2);

%for different initial phase angle, wt starts from -y axis
X6=-lambda*x5-lambda*x1*cos(x2)-1/wd^2-delta*cos(t);

yprime=[X1;X2;X3;X4;X5;X6];
```

E2.2   Function kshaft.m

```
function [FC,WK]=kshaft(R,r,C,B,m)
%function [k,K,FC,WK]=kshaft(R,r,C,B,m)
l=20.75*25.4/1000; %Span length
lc=5*25.4/1000; % lc is the length of the center weight
Rc=3/2*25.4/1000;% Rc is the O.D. of the center weight

mreal=2*m; %m:the weight of whole rotor as a reference only
to check whether we get the correct mass calculation.
elas=30*10.^6*0.4536/0.0254^2*10;
```

```
kesai=C/R;
g=9.81;1

%calculate m1 & m2 according to volume*density
al=(l-lc)/2;% al is the length "a"
bl=lc/2;% bl is the length "b"
Rmiddle=Rc;
volends=(al*2)*pi*(R^2-r^2);
volmiddle=(bl*2)*pi*(Rmiddle^2-r^2);
m1=volends*7800;
m2=volmiddle*7800;
m=m1+m2;
massdeviationpercentage=(m-mreal)/mreal*100;

% calculate the deflection in the center of the shaft
I=pi*(R^4-r^4)/4; % I:I1
Ic=pi*Rc^4/4; % Ic:I2
ww1=m1*g/(2*al);%w1
ww2=m2*g/(2*bl);%w2
AA1=-1/(elas*I); %A1
AA2=-1/(elas*Ic); %A2
F=AA2*(al+bl)^2/24*(2*ww2*al*bl+ww2*bl^2+2*ww1*al^2-
ww2*al^2)+AA1/24*(3*ww1*al^4-8*(ww1*al+ww2*bl)*al^3)
+AA2/24*(8*ww2*al^3*bl+6*ww1*al^4-ww2*al^4); %The
deflection in the center of the shaft
FC=F/C;
k=(m/2*g)/F;
K=k*C/(m/2*g);
WK=sqrt(k/(m/2));
```

Appendix F

Nomenclature

b_{ij} the linearized damping coefficients in polar coordinates, $N \cdot s/m$, $(i, j = \varepsilon, \phi)$ (i is the direction of the force and j is the direction of the velocity)

$$\bar{b}_{ij} = \frac{(C/R)^3}{\mu L} b_{ij}, \ (i, j = \varepsilon, \phi)$$

$$\bar{B}_{ij} = \frac{\omega C}{W} b_{ij}, \ (i, j = \varepsilon, \phi)$$

C radial clearance, m

D journal diameter, m

d_{eq} the linearized equivalent damping coefficient, $N \cdot s/m$

$$\bar{D}_{eq} = \frac{C\omega}{W} d_{eq}$$

d_{IJ} bearing damping coefficients in Cartesian coordinates, $N \cdot s/m$. $(I, J = x, y)$ (I is the direction of the force and J is the direction of the velocity)

$$\bar{D}_{IJ} = \frac{C\omega}{W} d_{IJ}$$

f_ε radial pressure force component, N

$\bar{f}_\varepsilon$ dimensionless radial pressure force component, $\bar{f}_\varepsilon = \dfrac{f_\varepsilon}{mC\omega^2}$

Thermohydrodynamic Instability in Fluid-Film Bearings, First Edition.
J. K. Wang and M. M. Khonsari.
© 2016 John Wiley & Sons, Ltd. Published 2016 by John Wiley & Sons, Ltd.

f_ϕ tangential pressure force component, N

$\bar{f}_\phi$ dimensionless tangential pressure force component, $\bar{f}_\phi = \dfrac{f_\phi}{mC\omega^2}$

F_ε radial drag force component, N

$\bar{F}_\varepsilon$ dimensionless radial drag force component, $\bar{F}_\varepsilon = \dfrac{F_\varepsilon}{mC\omega^2}$

F_ϕ tangential drag force component, N

$\bar{F}_\phi$ dimensionless tangential drag force component, $\bar{F}_\phi = \dfrac{F_\phi}{mC\omega^2}$

g gravitational constant, m/s^2

G_z turbulent coefficient

G_θ turbulent coefficient

k half-rotor stiffness, N/m

$\bar{K}$ dimensionless half-rotor stiffness

k_{eq} bearing equivalent stiffness coefficient, N/m

$$\bar{K}_{eq} = \frac{C}{W}k_{eq}$$

k_{ij} the linearized stiffness coefficients in polar coordinates, N/m, $(i,j=\varepsilon,\phi)$ (i is the direction of the force and j is the direction of the displacement)

$$\bar{k}_{ij} = \frac{(C/R)^3}{\mu\omega L}k_{ij}, \ (i,j=\varepsilon,\phi)$$

$$\bar{K}_{ij} = \frac{C}{W}k_{ij}, \ (i,j=\varepsilon,\phi)$$

k_{IJ} the linearized stiffness coefficients in Cartesian coordinates, N/m. $(I,J=x,y)$ (I is the direction of the force and J is the direction of the displacement)

$$\bar{K}_{IJ} = \frac{C}{W}k_{IJ}, \ (I,J=x,y)$$

k_r half-rotor stiffness coefficient, N/m

$$\bar{K}_r = \frac{C}{W}k_r$$

k_s effective stiffness coefficient of the rotor-bearing system, N/m

l rotor span, m

L bearing length, m

m rotor mass per bearing, kg

mu shaft unbalance per bearing, kg · m

O_b center of the journal bearing

O_j center of the journal of the shaft

O_{js} steady-state equilibrium position of journal center
O_{jd} dynamic position of journal center
O_{dg} geometric center of rotor central disk
O_{dm} mass center of rotor central disk
p_i inlet pressure, Pa
R journal radius, m
Re Reynolds number, $Re = \rho R \omega C / \mu$
R_s Stability envelope
s subscript represents the steady-state equilibrium condition

S Sommerfeld number, $S = \dfrac{\mu R L \omega}{\pi m g} \left(\dfrac{R}{C} \right)^2$

S_m modified Sommerfeld number, $S_m = \dfrac{\mu \omega R L^3}{2 m g C^2} = 2\pi (L/D)^2 S$

$S_p(\bar{\omega})$ the characteristic exponent determining the stability of the periodic solution of journal orbit at speed $\bar{\omega}$
t time, s
$T(\bar{\omega})$ period of the periodic solution of journal orbit at speed $\bar{\omega}$
T_{in} inlet temperature, °C
$\mathbf{v}_1$ an eigenvector of the Jacobian matrix at the stationary point for the critical value ν_c
ν_c critical value of Hopf bifurcation
u the mass eccentricity of rotor central disk
W load per bearing, N
x the coordinate in the horizontal direction
$\mathbf{x}_s$ the static equilibrium position
y the coordinate in the vertical direction
z the coordinate in axial direction
β_2 the leading coefficient in the expansion of a characteristic exponent
δ_{mid} the static central deflection of a flexible rotor, m
ε eccentricity ratio
ϕ attitude angle
γ_2 the number that gives the direction of the Hopf bifurcation

Γ bearing's characteristic number, $\Gamma = \dfrac{\mu R L^3}{2 m C^{2.5} g^{0.5}}$.

μ lubricant viscosity, Pa · s
ω running speed of the rotor, rad/s
$\bar{\omega}$ dimensionless running speed of the rotor, $\bar{\omega} = \omega \sqrt{C/g}$.
ω_c first critical speed of the rotor-bearing system, rad/s
ω_n the first natural frequency of the simply supported rotor system, rad/s
$\bar{\omega}_n$ dimensionless natural frequency of a simply supported rotor, $\bar{\omega}_n = \omega_n \sqrt{C/g}$

ω_{nd} undamped natural frequency of the system, rad/s

ω_s Threshold whirling frequency of the subsynchronous vibration, rad/s

$\bar{\omega}_s$ dimensionless whirl frequency at the threshold speed ω_{st}, $\bar{\omega}_s = \omega_s\sqrt{C/g}$.

ω_{st} threshold speed of the system, rad/s

$\bar{\omega}_{st}$ dimensionless threshold speed of the system, $\bar{\omega}_{st} = \omega_{st}\sqrt{C/g}$.

$\omega_0 = \beta(\omega_{st})$

Ω whirl frequency ratio at the threshold speed ω_{st}, $\Omega = \omega_s/\omega_{st}$

τ_2 the coefficient in the expansion of the periods of periodic solutions

Index

absolute circumferential coordinate, 13, 22, 23, 26, 27, 34, 135, 137, 139
ambient pressure, 6
asymptotic phase, 61
attitude angle, 1, 9, 11–13, 19, 20, 25, 26, 29, 33, 65, 70, 109, 130, 145, 149, 175, 195
Automated Diagnostics for Rotating Equipment (ADRE), 77

bearing equivalent damping, 92
bearing equivalent stiffness, 92, 94, 95, 194
bearing's characteristic number, 115, 118, 119, 124, 129, 195
bifurcation parameters, 61, 68, 70, 71, 83, 85, 156, 157, 163, 164
bisection method, 65, 70
Bode plot, 78–80, 97

cavitation, 6, 13, 14, 16–20, 22, 23, 25, 26, 34–37, 56, 107, 133, 162, 166, 171, 176

clearance circle, 3, 4, 29, 30, 40, 43, 64, 65, 69–74, 79–82, 87, 159, 160, 188
computerized data acquisition system, 77, 97

damping coefficients, 41–44, 46–48, 57, 94, 95, 110, 111, 147, 150, 152–155, 193
dip phenomenon, 63, 92, 97, 98
drag force, 63, 89, 91, 160, 162–165, 167, 194

eddy current proximity probes, 77
effective half-rotor stiffness, 92
elliptical journal bearing, 138
equations of motion, 30, 31, 34, 35, 37, 40, 48, 51, 55, 60, 61, 66, 67, 69, 70, 72, 82, 106, 107, 112–115, 117, 120, 126, 141–144, 147–150, 156, 160, 162–165, 179, 180
existence of the periodic solutions, 61, 68
external perturbation, 25, 64, 125

Thermohydrodynamic Instability in Fluid-Film Bearings, First Edition.
J. K. Wang and M. M. Khonsari.
© 2016 John Wiley & Sons, Ltd. Published 2016 by John Wiley & Sons, Ltd.